Intelligent Control and Learning Systems

Volume 14

Series Editor

The Springer book series Intelligent Control and Learning Systems addresses the emerging advances in intelligent control and learning systems from both mathematical theory and engineering application perspectives. It is a series of monographs and contributed volumes focusing on the in-depth exploration of learning theory in control such as iterative learning, machine learning, deep learning, and others sharing the learning concept, and their corresponding intelligent system frameworks in engineering applications. This series is featured by the comprehensive understanding and practical application of learning mechanisms. This book series involves applications in industrial engineering, control engineering, and material engineering, etc.

The Intelligent Control and Learning System book series promotes the exchange of emerging theory and technology of intelligent control and learning systems between academia and industry. It aims to provide a timely reflection of the advances in intelligent control and learning systems. This book series is distinguished by the combination of the system theory and emerging topics such as machine learning, artificial intelligence, and big data. As a collection, this book series provides valuable resources to a wide audience in academia, the engineering research community, industry and anyone else looking to expand their knowledge in intelligent control and learning systems.

Juan Zhang · Huaguang Zhang · Dongsheng Yang ·
Yanhong Luo · Bowen Zhou

Adaptive Collaborative Control of Multiagent Systems by Event-Triggered Mechanisms

Springer

Juan Zhang
Northeastern University
Shenyang, Liaoning, China

Huaguang Zhang
Northeastern University
Shenyang, Liaoning, China

Dongsheng Yang
Northeastern University
Shenyang, Liaoning, China

Yanhong Luo
Northeastern University
Shenyang, Liaoning, China

Bowen Zhou
Northeastern University
Shenyang, Liaoning, China

ISSN 2662-5458 ISSN 2662-5466 (electronic)
Intelligent Control and Learning Systems
ISBN 978-981-97-7865-2 ISBN 978-981-97-7866-9 (eBook)
https://doi.org/10.1007/978-981-97-7866-9

This Springer imprint is published by the registered company Springer Nature Singapore Pte Ltd.
The registered company address is: 152 Beach Road, #21-01/04 Gateway East, Singapore 189721, Singapore

Preface

With the rapid development of network communication technology and artificial intelligence technology, as well as the deepening of the complexity of system control tasks, distributed collaborative control of multiagent systems has become one of the research directions that receive much attention in the field of control and is widely used in satellite formations, sensor networks, drones, and other fields. Collaborative control mainly includes consistency problems, output regulation problems, inclusion problems, formation problems, etc. However, the design of most existing control protocols relies on the global information of the communication topology. However, for multiagent systems with large-scale networks, it is difficult to obtain the eigenvalues of the Laplacian matrix of their communication topology. Therefore, considering the distributed collaborative control problem of multiagent systems has important research value and significance.

At present, although the research on adaptive cooperative control problems of multiagent systems has achieved fruitful results, there are still some theoretical problems to be solved. For example, practical systems are usually equipped with limited communication and computation resources, and continuous communication and/or continuous controller updates will greatly shorten the service life of the system. With the increasing scale of the constituent multiagent systems, it is difficult for each agent to obtain the topology of the communication network in real time, and it is not possible to compute the eigenvalues of the Laplace matrix in real time, so the control laws that rely on global information about the communication topology graph are impractical. In addition, multiagent systems face many security problems, especially denial-of-service (DoS) attacks at the network layer and actuator faults at the physical layer, where DoS attacks block information transmission in the communication network and actuator faults lead to system performance degradation, which affects the stability of the system. In this book, based on the summary of the previous work, we give the corresponding solutions to the above problems. The relevant theoretical results are applied to the DC motor-driven single-link manipulator model with rotary joints and an inverted pendulum model. The theoretical results are applied to a DC motor-driven single-link manipulator model with a rotary joint and an inverted pendulum model. The obtained results demonstrate the effectiveness of the proposed method. The specific research and main contribution in this paper include the following six aspects:

1. On the one hand, the leaderless and leader-follower consensus problems of homogeneous linear multiagent systems are investigated for the switching undirected communication topology graph. The novel model-based event-triggered observers are proposed to observe the state of each system, and the distributed adaptive event-triggered consensus protocols are designed based on the output information, where the dimension of the information exchanged among the neighbors is the same as that of the system's output, i.e., the proposed control protocols require a lighter communication load. Therefore, the proposed method can effectively alleviate the pressure of agent information transmission without increasing the network bandwidth. On the other hand, the leader-follower consensus problem of homogeneous linear multiagent systems is investigated for directed communication topology graphs with communication link faults. An adaptive hybrid event-triggered method is used to solve the leader-follower consensus problem of multiagent systems under the directed topology graph with communication link faults. The proposed method avoids continuous communication among neighbors, saves the communication resources of the system, and extends the service life of the multiagent systems.
2. For the undirected communication topology graph, the consensus problems of nonlinear multiagent systems with Lipschitz dynamics and one-sided Lipschitz dynamics are investigated, respectively. Firstly, node-based and edge-based event-triggered mechanisms are designed for nonlinear multiagent systems with Lipschitz condition, respectively, and static and dynamic distributed consensus protocols are proposed, where the latter is designed independently of any global information of the communication topology graph. The proposed method realizes asynchronous event-triggered communication of neighbors, which effectively avoids the transmission of unnecessary information and thus saves communication resources. Secondly, the node-based event-triggered consensus protocol is proposed for nonlinear multiagent systems with one-sided Lipschitz dynamics, which solves the leader-follower consensus problem and the leaderless consensus problem of nonlinear multiagent systems with one-sided Lipschitz dynamics, respectively. The proposed node-based event-triggered mechanism can reduce the waste of system communication resources. Finally, two simulation examples further validate the effectiveness of the proposed method.
3. For heterogeneous linear multiagent systems with DoS attacks and actuator faults, the output regulation problem is investigated, in which the leader system matrix is only available to the subsystems directly connected to the leader to obtain the system matrix and state of the leader. For this purpose, the edge-based event-triggered observer is first designed to estimate the leader's system matrix and state under the influence of DoS attack, based on which the distributed adaptive event-triggered fault-tolerant controller is proposed to solve the output regulation problem of the heterogeneous linear multiagent system under the influence of actuator faults to overcome the influence of actuator faults on the stability of the

multiagent systems. And the edge-based event-triggered mechanism and node-based event-triggered mechanism are designed, while the asynchronous event-triggered communication between neighbors and the event-triggered update of the controller are realized. The proposed method can not only guarantee the network to operate safely and maximize the loss recovery when hackers attack, but also the multiagent systems can still operate stably when the actuator fails.

4. Considering the containment problem of homogeneous linear multiagent systems under directed topology graphs, due to the asymmetry of the Laplace matrix, the existing adaptive mechanism methods for dealing with undirected communication topology graphs are no longer feasible relying on the symmetry property of the Laplace matrix and are not applicable to the case of directed topology graphs. For the multiagent system with zero input leader, firstly, a distributed adaptive event-triggered containment control protocol and composite event-triggered mechanism are constructed based on relative information. Secondly, to avoid continuously supervising the information of neighbors, a new computational method is proposed to ensure that the agents measure the relative information only at the triggering instants. Finally, the proposed method is extended to the case with a non-zero input leader by adding an extra term in the triggering mechanism and introducing an extra nonlinear term in the controller compared to the case with the zero input leader. In addition, numerical and comparative simulations are introduced to illustrate the advantages of the proposed method in terms of fast convergence, low communication frequency of the agents, and low update frequency of the controller.
5. The formation tracking problem of heterogeneous linear multiagent systems with disturbance is studied for undirected topology graphs. The goal is to make the outputs of all the followers complete a preset formation configuration and simultaneously track a convex combination consisting of the outputs of the leader. A distributed dynamic event-triggered compensator is first designed to estimate the convex combinations of the leader's states, and in order to further expand the triggering intervals, internal variables are introduced into the dynamic event-triggered mechanism to dynamically adjust the threshold of each follower. Compared with the traditional static threshold, the time-varying threshold ensures a larger neighbor triggering time interval. Then, an output formation control protocol based on a distributed adaptive event-triggered compensator, a compensation input, and a time-varying formation signal is proposed, where the compensation input provides a degree of freedom for the feasible formation set. Finally, the main results are verified by two examples. The proposed triggering mechanism enables the agents to satisfy discrete-time measurement and discrete-time communication and gives the lower bound of the triggering time interval of adjacent events.
6. The formation containment problem for heterogeneous linear multiagent systems is studied for directed communication topology graphs, where the dynamics of the leaders are heterogeneous and are partitioned into virtual leaders, informed formation leaders, and uninformed formation leaders. Firstly, the adaptive observers are designed to estimate the states of the virtual leaders for the

informed and uninformed formation leaders, respectively. Secondly, the formation control protocol is designed for the formation leaders to enable the formation leaders to form the established formation. Finally, the containment control protocol is designed so that the followers eventually enter into the formation formed by the formation leader, solving the formation containment control problem for heterogeneous multiagent systems. By introducing a dynamic formation function, collisions between agents and obstacles can be avoided, and the formation coverage area can be automatically adjusted according to the control requirements. Therefore, the proposed method has a wider range of applications.

Shenyang, China

Juan Zhang
Huaguang Zhang
Dongsheng Yang
Yanhong Luo
Bowen Zhou

Acknowledgements This book was supported by National Natural Science Foundation of China under 62303096, China Postdoctoral Science Foundation under 2023M740544, 2023 National Postdoctoral Research Program (Level C) under GZC20230389, Joint Funds of the National Natural Science Foundation of Liaoning under 2023-MSBA-105, and Fundamental Research Funds for the Central Universities under N2404020.

Contents

About the Authors

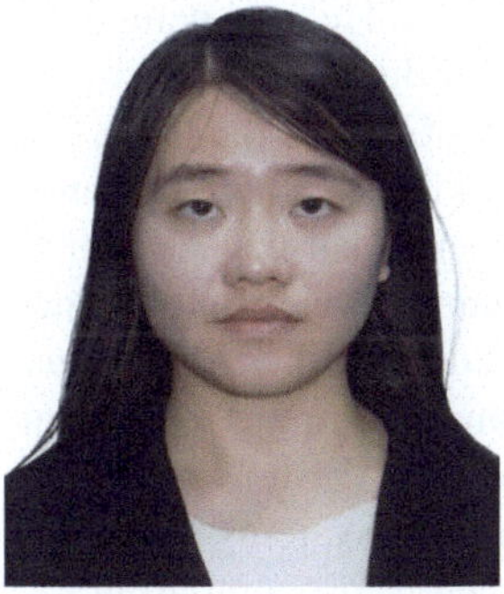

Juan Zhang (Member, IEEE) received the B.S. degree in information and computing science from Shanxi Normal University, Linfen, China, in 2016, and the M.S. degree in basic mathematics and the Ph.D. degree in control theory and control engineering from Northeastern University, Shenyang, China, in 2018 and 2023, respectively. She is currently a Lecturer at the School of Information Science and Engineering, Northeastern University, where she is also a Postdoctoral Fellow. Her current research interests include containment control, event-triggered control, formation control, consensus control, fault-tolerant control, DoS attacks for multiagent systems, and their applications in boiler systems.

Author Details

Juan Zhang (张娟)
Lecture
Northeastern University
Boston, USA
Homepage: http://faculty.neu.edu.cn/zhangjuan/zh_CN/index.htm e-mail: zhangjuan@ise.neu.edu.cn

Huaguang Zhang (Fellow, IEEE) received the B.S. and M.S. degrees in control engineering from Northeast Dianli University, Jilin, China, in 1982 and 1985, respectively, and the Ph.D. degree in thermal power engineering and automation from Southeast University, Nanjing, China, in 1991. He has authored or coauthored over 280 journals and conference 897 papers, six monographs, and co-invented 90 patents. His current research interests include fuzzy control, stochastic system control, neural networks-based control, nonlinear control, and their applications. He was a recipient of the Outstanding Youth Science Foundation Award from the National Natural Science Foundation Committee of China in 2003 and the IEEE Transactions on Neural Networks and Learning Systems Outstanding Paper Award in 2012. He was named the Cheung Kong Scholar by the Education Ministry of China in 2005. He is the E-Letter Chair of the IEEE CIS Society and the Former Chair of the Adaptive Dynamic Programming and Reinforcement Learning Technical Committee of the IEEE Computational Intelligence Society. He was an Associate Editor of the IEEE Transactions on Fuzzy Systems from 2008 to 2013. He is an Associate Editor of Automatica, the IEEE Transactions on Neural Networks and Learning Systems, the IEEE Transactions on Cybernetics, and Neurocomputing.

Author Details

Huaguang Zhang (张化光)
Professor
Northeastern University
Boston, USA
Homepage: http://faculty.neu.edu.cn/zhanghuaguang/zh_CN/index.htm e-mail: zhanghuaguang@mail.neu.edu.cn

Dongsheng Yang (Senior Member, IEEE) received the B.S. degree in testing technology and instrumentation, the M.S. degree in power electronics and electric drives, and the Ph.D. degree in control theory and control engineering from Northeastern University, Shenyang, China, in 1999, 2004, and 2007, respectively. He is currently a Professor at Northeastern University. He was supported by the Program for New Century Excellent Talents at the University. He has authored or coauthored more than 70 papers published in academic journals and conference proceedings, three monographs, and co-invented 80 patents. His current research interests include distributed generation, multienergy power systems, and artificial intelligence-based fault diagnosis and protection. He was a recipient of the Second Prize of National Science and Technology Progress.

Author Details

Dongsheng Yang (杨东升)
Professor
Northeastern University
Boston, USA
Homepage: http://faculty.neu.edu.cn/yangdongsheng/zh_CN/index.htm e-mail: yangdongsheng@mail.neu.edu.cn

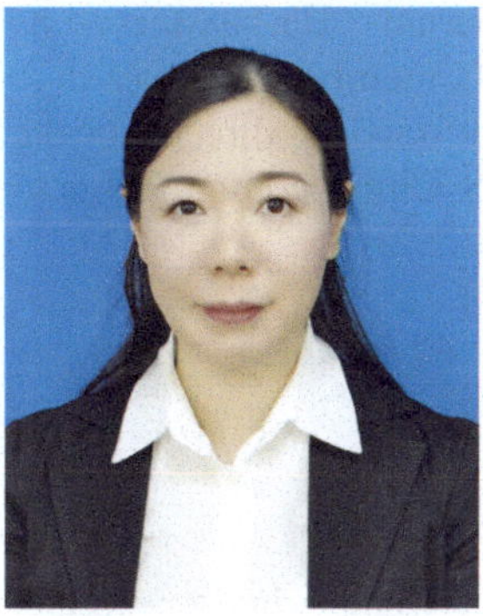

Yanhong Luo (Member, IEEE) received the B.S. degree in automation control and the Ph.D. degree in control theory and control engineering from Northeastern University, Shenyang, China, in 2003 and 2009, respectively. From 2016 to 2017, she visited The University of Texas at Arlington as a Visiting Scholar. She is currently a Professor at Northeastern University. She has published more than 100 SCI/EI-indexed papers. Her research interests include renewable energy accommodation, distributed control and optimal operation of energy Internet, adaptive dynamic programming, and data-driven self-learning control and its industrial applications. She is a recipient of the Andrew P. Sage Best Transactions Paper Award (2015) with the IEEE SMC Society. In addition, she is the Former Vice-Chair of the Adaptive Dynamic Programming Reinforcement Learning Technical Committee with IEEE Computational Intelligence Society. She served as the Session

Chair or a TPC member for more than 20 international conferences.

Author Details
Yanhong Luo (罗艳红)
Professor
Northeastern University
Boston, USA
Homepage: http://faculty.neu.edu.cn/luoyanhong/zh_CN/index.htm e-mail: luoyanhong@ise.neu.edu.cn

Bowen Zhou (Member, IEEE) received the B.Sc. and M.Sc. degrees in electrical engineering from Wuhan University, Wuhan, China, in 2010 and 2012, respectively, and the Ph.D. degree in electrical engineering from Queen's University Belfast, Belfast UK, in 2016. He joined the Department of Electrical Engineering, Northeastern University, Shenyang, China, in 2016, where he is currently an Associate Professor. He is also the Vice Dean of the School of Future Technologies, Northeastern University. He has published more than 70 SCI or EI-indexed papers and holds 20 patents. His research interests include power and energy system operation, energy storage, demand response, and energy Internet. He has served as a youth editorial board member and a special issue guest editor for several international and domestic journals. He has served as the session chair and a TC/PC member for more than ten conferences. He is also the standing director or the director of several IEEE PES China committees and subcommittees.

Author Details
Bowen Zhou (周博文)
Associate Professor
Northeastern University
Boston, USA
Homepage: http://faculty.neu.edu.cn/zhoubowen/zh_CN/index.htm e-mail: zhoubowen@ise.neu.edu.cn

Chapter 1
Introduction

1.1 Background of Multiagent Systems

Agent generally refers to a computing entity that resides in a certain environment, can continuously and autonomously play a role, and has characteristics such as residency, responsiveness, sociality, and proactivity. Any entity that can interact with the environment can be abstracted as an agent. A single agent typically exhibits the most basic characteristics of autonomy, social ability, responsiveness, foresight, and rationality. Since John McCarthy proposed the concept of "artificial intelligence" in 1956, the concept of agents has been on the rise. Before the 1970s, agents were almost always studied as individual entities. In 1978, the US Department of Defense focused on building a multiagent system at a distributed sensor network seminar held at Carnegie Mellon University. A multiagent system is a collection of multiple agents that communicate and collaborate with each other, as well as with the environment, to jointly solve problems that a single agent cannot solve. Subsequently, multiagent systems are widely discussed in different branches of the field.

There are various ways to construct multiagent systems. According to the dynamics of individual agents that make up multiagent systems, multiagent systems can be divided into two types: homogeneous and heterogeneous multiagent systems. In a combat scenario, homogeneous multiagent system refers to each army having only one type of combat unit (such as a Marine Corps), and heterogeneous multiagent system refers to the presence of multiple combat units in one party's agent. Specifically, homogeneous multiagent systems can be seen as a special case of heterogeneous multiagent systems. According to the control methods of multiagent systems, they can be divided into three types: centralized, decentralized, and distributed. Centralized control refers to the centralized control of the entire system by one agent, which is a top-down hierarchical control structure for planning and decision-making. The system has good coordination, but poor real-time and dynamic performance, and poor responsiveness to environmental changes; Decentralized control refers to the high degree of autonomy of each agent, which processes information, plans and makes decisions, executes instructions on its own, and communicates with other agents to

J. Zhang et al., *Adaptive Collaborative Control of Multiagent Systems by Event-Triggered Mechanisms*, Intelligent Control and Learning Systems 14,
https://doi.org/10.1007/978-981-97-7866-9_1

coordinate their respective behaviors, without a centralized control unit. Distributed structures have good fault tolerance and scalability, but require high communication requirements and low efficiency in multilateral cooperation, making it difficult to ensure the achievement of global goals; Distributed control, whose structure lies between the above two structures, is a hierarchical and locally centralized structure where all agents are globally equivalent. This structure combines decentralized horizontal interaction with centralized vertical control, which not only improves coordination efficiency but also does not affect the real-time, dynamic, and fault tolerance of the system. Adaptive distributed control refers to algorithms designed on the basis of distributed control that do not rely on the entire communication topology. Compared with the previous control methods, adaptive distributed control has brought significant advantages in terms of computational complexity, communication cost, system robustness, and scalability.

Inspired by the collaborative behavior of biological groups such as fish gathering and bird migration, the collaborative control of multiagent systems has been widely applied in engineering fields, such as satellite formation in Wu et al. (2012), sensor networks in Rogers et al. (2009), unmanned aerial vehicles in Zhou et al. (2022), and so on, and has become one of the research directions that have received much attention in the field of control. The application of multiagent system collaborative control is reflected in different aspects, such as military, transportation control, intelligent robots, and so on. In terms of military affairs, the authors in Reynolds (1987) utilize the multiagent technology to construct a firepower allocation model, achieving the intelligence of the firepower model. The authors in Li (2004) have established a dialogue mode for the communication between tank intelligent agents, providing effective reference value for the simulation development of tank units. In terms of transportation control, Vicsek et al. in (1995) use multiagent coordinated control to coordinate control signals at adjacent intersections, in order to eliminate traffic network congestion. In terms of intelligent robots, the RoboCup football World Cup hosted by robots in Japan is a typical example. A single robot cannot complete a task, they must work together to complete the task.

In the above context, this book considers the distributed collaborative control problem of multiagent systems. For multiagent systems with large-scale networks, this paper proposes several adaptive distributed control protocols to solve the collaborative control problem of multiagent systems. The control problems considered in this book have important theoretical significance and practical application value.

1.2 Research Status and Progress

The research of multiagent system mainly includes consensus, output regulation, containment, and formation problem. However, multiagent technology has also encountered some bottlenecks, such as the high communication cost caused by the introduction of a large number of agents in the process of modeling complex systems, and the extreme limitations of the system's computing and storage resources. Ensuring normal collaboration between agents is also extremely challenging. Especially for multiagent systems with large-scale networks, their communication topology is vast, and it is unrealistic for the design of controller to rely on the global information of the communication topology. Therefore, adaptive control, event-triggered control, and edge-based event-triggered control have been introduced to address these issues. Below is a brief introduction to the current research status of these issues both domestically and internationally.

1.2.1 Consensus Problem

Consensus is a fundamental issue in the research of collaborative control in multiagent systems. Based on the existence of leaders in multiagent systems, consensus problems are divided into leaderless consensus and leader-follower consensus problems. Leaderless consensus refers to designing appropriate distributed consensus protocols to enable a group of agents connected by communication networks to ultimately converge towards the same goal. Leader-follower consensus refers to designing appropriate distributed consensus protocols for each follower to drive all followers to ultimately track the leader's trajectory.

The formal research of scientists on consensus issues originates from management science and statistics in the 1960s. Distributed computing on the network has a tradition in systems and control theory, which begins with pioneering work on asynchronous asymptotic consensus problems in distributed decision systems and asynchronous asymptotic consensus problems in parallel computing. On the basis of early work in Fax and Murray (2004), Olfati Saber et al. in Olfati-Saber and Murray (2004) propose a theoretical framework for solving the consensus problem of first-order network dynamic systems in 2004, develop consensus algorithms, and analyze theoretically that the consensus error of multiagent systems can converge to zero under the conditions of undirected communication topology and strongly connected balanced topology. Finally, based on the knowledge of algebraic graph theory, the necessary and sufficient conditions for achieving average consistency in multiagent systems with topological graphs satisfying equilibrium graph conditions are provided. Ren and Beard (2015) studied the consensus problem of multiagent systems with discrete and continuous dynamics in 2005, respectively. The proposed method is applicable to the case of weighted directed graphs and pointed out that the condition for achieving consensus in multiagent systems is that the communication topology includes a spanning tree. Subsequently, researchers from various disciplines such as engineering and science became extremely interested in the issue of consensus in multiagent systems and designed consensus protocols for various multiagent system models. For example, Liu et al. in (2017, 2018) study the average consensus problem of first-order and second-order multiagent systems, respectively. Ren et al. in (2006) analyze multiagent systems with high-order linear integrator models and obtain necessary and sufficient conditions to solve consensus problems, laying a theoretical foundation for the study of consensus in high-order linear multiagent systems. The authors in Yu et al. (2011a) study the leader-follower consensus problem for high-order multiagent systems and provide necessary conditions for high-order consensus. The authors in Ni and Cheng (2010) consider the leader-follower consensus problem of linear multiagent systems under fixed communication topology and switched communication topology, using techniques such as algebraic graph theory, Riccati inequality, and Lyapunov inequality. The authors in Rehan et al. (2018) study the consensus problem of the Lipschitz nonlinear multiagent system based on relative state feedback by designing a distributed consistency control law.

Note that some of the references mentioned rely on the global information of the communication topology, such as the eigenvalues of the Laplacian matrix and the number of multiagent systems. However, with the deepening application of multiagent systems in various fields, modeling complex systems may introduce a large number of agents. In this case, it is not easy for each agent to obtain the entire communication topology map. To address these issues, inspired by the complex network synchronization adaptive strategy (DeLellis et al. 2010), an adaptive mechanism (Li et al. 2012, 2015; Lv et al. 2016) is introduced into the design of the control law to adjust the coupling weights between adjacent agents instead of fixed weights. For example, Li et al. (2012) separately study the leaderless consensus and leader-follower consensus problems of multiagent systems with general linear Lipschitz nonlinear dynamics under undirected communication topologies. The application of distributed adaptive coupling weights makes the design of control protocols independent of any global information. Due to the limited resources that may exist in actual systems, including communication resources, computing resources, and so on, designing a distributed event-triggered consensus protocol (Zhang et al. 2014; Hu et al. 2016) is crucial. Two event-triggered mechanism schemes, namely centralized and distributed schemes, have been proposed in Zhang et al. (2014) to solve the problem of leaderless consensus of linear multiagent systems. The centralized triggering control scheme is suitable for small-scale multiagent systems. The design of centralized and distributed triggering mechanisms in Zhang et al. (2014) and Hu et al. (2016) relied on the global information of the communication topology and is not suitable for multiagent systems with large-scale networks. To address these issues, the distributed adaptive event-triggered consensus protocols in Li et al. (2020), Cheng and Li (2018), ad Qian et al. (2019) are introduced to address the consensus problem of multiagent systems. This method is suitable for multiagent systems with large-scale networks.

Based on the above discussion, designing the distributed adaptive event-triggered control scheme to solve the consensus problem of multiagent systems is of great application value and practical significance. The existing results on consensus research only ensure that agent achieves intermittent communication and cannot guarantee the event-triggered update of controllers. It is worth mentioning that the dual event-triggered mechanism in Jiang et al. (2022) can ensure both intermittent communication and intermittent controller updates between neighbor agents. However, this method requires high hardware requirements for agents. Therefore, how to design a distributed adaptive event-triggered control protocol different from the dual event-triggered scheme to ensure intermittent communication of agents and update of intermittent controllers is still a problem that needs to be considered.

1.2.2 Output Regulation Problem

The output regulation problem, also known as the servomechanism problem, has been one of the fundamental problems in control theory since the 1970s. It aims at designing a feedback controller such that asymptotic tracking and disturbance rejection can be achieved while maintaining the stability of the closed-loop control system. This is a general formulation for many classic control problems, such as vibration suppression of high-speed trains, manipulation of robot arms, tracking of a moving target by a missile, and controlling of a spacecraft subject to disturbances, to name just a few. Note that the output regulation problem can be seen as an extension of the leader-follower consensus problem.

Output regulation problem has attracted the attention of many scholars and also has promoted the development of modern control theory and applications. Up to now, considerable research results have been achieved in Su and Huang (2012b), Li et al. (2016), Yang et al. (2018), and reference therein. For example, Su and Huang (2012b) investigate the collaborative output regulation problem of linear multiagent systems under switching topologies, where continuous-time state feedback and output feedback control protocols are designed, respectively. In order to remove the dependence on the global information of the communication topology, Li et al. in (2016) design the distributed adaptive controller to solve the output regulation problem of linear multiagent systems with directed communication topology, where the designed distributed adaptive observer is used to estimate external signals. Yang et al. in Yang et al. (2018) investigate the collaborative output regulation problem of heterogeneous linear multiagent systems under switching topology, where the distributed adaptive event-triggered control strategies based on state feedback and output feedback are proposed. The methods avoid the use of the minimum non-zero eigenvalue of the Laplacian matrix related to the communication topology and achieve the event-triggered communication of agents. The above references assume that each subsystem knows the conditions for the external system matrix. Cai et al. in (2017) point out that this may not be practical in certain applications. For example, if the leader's signal is a sin signal, assuming that each follower knows the external system matrix is equivalent to assuming that each follower's control knows the frequency of the sin signal. Therefore, the distributed observer is first designed to estimate the external system matrix and state, and then the distributed controller is designed to solve the output regulation problem of the linear multiagent system with a direct communication topology based on the designed observer and the regulator equations solved online. However, the proposed method still relies on the global information and the agent needs continuous communication. The authors in Qian et al. (2021) make up for the shortcomings of the method proposed in Cai et al. (2017). Firstly, the adaptive event-triggered observer is designed to estimate the matrix and state of the external system, and then the distributed controller is designed to solve the output regulation problem of heterogeneous linear multiagent systems. However, the output regulation problem of multiagent systems discussed above is all under ideal network conditions, and the default actuator is not faulty. In fact, there may be DoS attacks during information transmission, and aging of equipment usage may cause actuator failures. Therefore, considering the output regulation problem of heterogeneous linear multiagent systems with the coexistence of DoS attacks and actuator faults is of great practical significance.

1.2.3 Containment Problem

Containment problem can be seen as an extension of the consensus problem. Due to its broad practical application background, such as residential protection, joint rescue, and so on, it is a hot topic in multiagent system cooperative control research. Containment control means that under the designed control protocol, all followers transmit local information through a communication network, and finally all followers enter the geometric space surrounded by all leaders.

Containment problem is first proposed in Dimarogonas et al. (2006), and the designed control algorithm is applied to the model of multiagent system composed of unicycle. Subsequently, relative research has attracted more and more attention from scholars, and many fruitful research results have emerged in Cao et al. (2012, 2011), Shi et al. (2017), Dimarogonas et al. (2009), Meng et al. (2010), Li et al. (2013), Ma et al. (2015), and Xiong et al. (2022). For example, the authors in Cao et al. (2012) consider the containment problem of first-order multiagent systems and propose an important coordinate transformation method to obtain the final convergence result, which ensures that all followers ultimately enter the convex hull composed of all leaders. Cao et al. in (2011) investigate the containment problem of a group of autonomous agents with multiple stationary/dynamic leaders under fixed and directed switching communication topologies. Li et al. in (2013) focus on the containment problem of continuous-time and discrete-time multiagent systems with general linear dynamics under a directed communication topology. Ma et al. in (2015) study the containment problem of linear discrete multiagent systems, in which sufficient and necessary conditions for implementing the containment problem are obtained using the knowledge of spectral radius analysis and matrix theory. However, the design of control protocols in the above-mentioned references depends on the global information of communication topology. In order to avoid relying on the global information for the communication topology, the distributed continuous-time controllers and adaptive continuous-time controllers are designed in Li et al. (2015) to solve the containment problem of linear multiagent systems with zero input leaders. In addition, in order to avoid continuous communication between neighbors, the distributed event-triggered control strategy is designed in Xiong et al. (2022) to solve the containment problem of multiagent systems, in which the designed event-triggered mechanism not only realizes event-triggered communication between neighbors, but also realizes intermittent update of controller. However, the dynamics of multiagent system is second-order, and the proposed method depends on the global information of communication topology. How to use adaptive mechanisms and event-triggered mechanisms to design control laws and solve the containment problem of linear multiagent systems is still an open problem.

1.2.4 Formation Problem

Formation control is the fundamental problem in cooperative control of multi-unmanned vehicle system and multi-unmanned aerial vehicle system, which means that in some special environments and tasks, agents need to achieve and maintain the present geometric structure under certain constraints and perform tasks in an efficient and reliable way. Up to now, many methods have been proposed to solve the formation problem, which are mainly summarized as leader-follower method in Li et al. (2018), behavior-based method in Song and Yang (2012), and virtual structure method in Liu and Qin (2021). The main advantages of the leader-follower method are strong expansibility, simple operation, and cost saving. At the same time, the leader-follower method also has some shortcomings, which are mainly reflected in the poor robustness of formation and extreme dependence on leaders. The advantages of behavior-based method are strong adaptability, the corresponding control scheme can be obtained according to the environment, and distributed control can be realized. Its disadvantage is that modeling and mathematical analysis are not easy to implement, and the stability of formation can not be guaranteed. The advantage of virtual structure method is that it has high precision trajectory tracking effect. Its disadvantage is that the formation is single, which limits its application scenarios. Note that the leader-follower method is the most common formation control method at present.

In the past decades, formation control of multiagent systems has attracted wide attention in various fields such as applications in robots, autonomous vehicles, underwater vehicles, and so on. Generally speaking, the formation of multiagent system can be divided into time-invariant formation Cheng et al. (2021) and time-varying formation (Dong and Hu 2017; Zuo et al. 2019; Hua et al. 2018). For example, the authors in Cheng et al. (2021) study the time-invariant formation problem of homogeneous linear multiagent systems and propose the edge-based event-triggered control scheme, which effectively avoids continuous communication between neighbors and reduces the high bandwidth demand for system communication and the system resource occupation. The authors in Dong and Hu (2017) study the time-varying formation tracking problem of homogeneous linear multiagent systems with multiple leaders, in which the followers form an ideal time-varying formation while tracking convex combination formed by leaders. The authors in Hua et al. (2018) study the time-varying output formation tracking problem of heterogeneous linear multiagent systems. The mentioned references consider the formation problem with one leader and multiple homogeneous leaders, but multiple homogeneous leaders can be equivalent to one leader. Moreover, the methods in Zuo et al. (2019) and Hua et al. (2018) depend on the eigenvalue information of Laplace matrix, which is global information and is not suitable for multiagent systems with large-scale networks. With the expansion of multiagent system modeling scale, communication networks have become more complex, and it is difficult for each agent to obtain global information. Therefore, the adaptive mechanism is introduced to solve the formation problem of multiagent systems. For example, the authors in Wang et al. (2019) solve the leaderless formation problem of linear multiagent systems under directed graphs using distributed adaptive output feedback control. The proposed method does not rely on the global information of the communication topology graph. The authors in Jiang et al. (2019) solve the formation problem of heterogeneous linear multiagent systems under directed communication topology, where a new time-varying formation condition is proposed. The authors in Hua et al. (2019) study the distributed adaptive time-varying formation tracking problem of heterogeneous linear multiagent systems with nonautonomous leaders under directed communication topology, where the leaders have unknown control inputs. In addition, in Zhang et al. (2022), to achieve intermittent asynchronous communication between agents, an adaptive edge-based event-triggered control scheme is designed, thereby realizing the time-varying formation tracking problem of linear multiagent systems with nonautonomous leaders. It is worth noting that the current consideration of formation problems mainly focuses on distributed formation tracking and formation containment problems. However, there are few research results on using distributed adaptive event-triggered schemes to solve the formation tracking and formation containment problems of heterogeneous linear multiagent systems under directed communication topologies. This is an area that remains to be studied.

1.2.5 Event-Triggered Control

In the cooperative control of multiagent systems, the agents need to exchange local information among themselves with the help of the communication topology to achieve the control goal. The traditional network sampling control method in Yu et al. (2011b) is a time-based periodic sampling scheme, where the triggering instants follow some predetermined pattern. Time-triggered control not only does not effectively reduce the waste of resources but also reduces the control efficiency of the system. In contrast, the sampling instants of event-triggered control in Tabuada (2007) are determined by a specific event, and sampling is performed when and only when the set event is satisfied. In practical applications, especially in wireless sensor network control, embedded UAV cooperative control, and other fields, there are problems such as relatively limited network bandwidth, computational resources, storage resources, etc., event-triggered control can better meet the needs of this type of applications than time-triggered control. It is worth mentioning that, although the event-triggered control strategy can effectively reduce the sampling frequency, if the event conditions are continuously satisfied, an infinite number of triggers will occur in a finite period of time, i.e., generating Zeno behavior. In this case, the designed event-triggered strategy fails and degrades to a continuous control situation, and the superiority of the method disappears.

Inspired by Tabuada (2007), the authors in Dimarogonas et al. (2012) propose both centralized and distributed event-triggered control schemes and demonstrate the average consensus of the multiagent system with the single integrator using Lyapunov stability theory, which shows that the proposed event-triggered scheme allows the agents to perform fewer control updates. Due to the unique advantages of event-triggered control, more and more scientists have utilized event-triggered control strategies to study the cooperative control problems of multiagent systems and have achieved fruitful research results in Hu et al. (2016), Ding et al. (2018), Seyboth et al. (2013), Zhu et al. (2014), Yang et al. (2016), Garcia and Antsaklis (2013), and Garcia et al. (2017). Common event-triggered control schemes are summarized into four types: event-based sampling schemes in Dimarogonas et al. (2012), model-based event sampling schemes in Garcia and Antsaklis (2013), event-triggered sampling schemes based on sampled data in Garcia et al. (2017), and self-triggered sampling schemes in Hu et al. (2016).

The distinctive feature of the event-based sampling scheme is that whether the current measurement is sampled or not is determined by a predefined triggering mechanism. This means that this strategy can significantly reduce the number of samples taken by the agent, which leads to significant savings in device energy consumption. A significant drawback of this sampling scheme in Hu et al. (2016) is the requirement for continuous communication, i.e., continuous communication between neighbors is necessary when determining the triggering mechanism. As a result, more communication resources are required to implement the event-based sampling scheme, which greatly increases the operational cost and is in contradiction. To avoid continuous communication of agents, a novel distributed event-triggered scheme is proposed in Yu and Antsaklis (2014), where the threshold in the triggering mechanism depends only on the discrete signals of the neighbors at the triggering instant. Another effective approach to avoid continuous communication is to use a simpler state-independent threshold for the relative error, as reported in Seyboth et al. (2013). In addition, one of the most common effective methods to avoid continuous communication is to design self-triggered sampling schemes.

The distinctive feature of the model-based event-triggered scheme is that the system state is estimated at the update interval. Based on the estimation error, the triggering mechanism is designed to determine when the actual system state should be transmitted. Two common approaches to system state estimation are open-loop estimation in Cheng and Li (2019a) and closed-loop estimation in Adaldo et al. (2015). The drawbacks of using open-loop estimation are summarized in three aspects. (1) Control updates are performed continuously, so it does not save the controller energy consumption of the agents, which is not favorable for multiagent systems with battery power. (2) It requires equipping the agents with additional estimators. The number of such estimators becomes larger as the number of neighbors of the agents increases, which leads to high implementation costs and high complexity. (3) The need for the system matrices to be identical, which restricts the scope of application of the event-triggered scheme. In general, the controller update information for each agent is considered private and may not be accessible to other agents. This will make it impossible to implement an event-triggered scheme in some practical applications. The use of closed-loop estimation requires the assumption that each agent has access to the control update values of its neighbors. Of course, the advantage of a model-based event-triggered scheme is that communication between agents occurs only at the triggering instants, which means that agents no longer need to communicate continuously.

The above-mentioned event-triggered schemes require constant checking and calculation of triggering conditions, which requires additional hardware support and increases the cost of system design, operation, and maintenance. To avoid these problems, the sampled data-based event-triggered scheme is proposed in Garcia et al. (2017). The core idea of the sampled data-based event-triggered control is that the event detection takes place only at the sampled instants and not continuously. Moreover, the minimum value of the time interval between adjacent events is essentially a sampling period, which means that the system must not exhibit Zeno behavior. However, its implementation may be subject to some limitations. For example, it is assumed that all agents are sampled synchronously within a constant sampling period. In practice, such sampling clock synchronization is difficult to achieve for large-scale networks. In addition, another scheme to avoid the continuous detection of event conditions is the self-triggered sampling scheme, which predicts the next sampling time at the instant of control update based on the latest triggered data and system dynamics information. However, the self-triggered mechanism exacerbates the communication frequency of the agents. Therefore, the user should make a trade-off in designing an event-triggered scheme to make the final choice.

The communication way between agents under the event-triggered mechanism mentioned in the above works is characterized by intermittent and synchronous communication. In fact, this event-triggered mechanism is also called node-based event-triggered mechanism. That is to say, at the triggering instant, the agent transmits its current information to all its outgoing neighbors simultaneously, even though there may be a situation in which the current information of one of the outgoing neighbors is similar or the same as that of the agent. However, in this case, the agent does not need to transmit information to its neighbors that are similar or identical to its information. In order to achieve this goal, edge-based event-triggered mechanisms in Cheng et al. (2021), Cheng and Li (2019b), and Meng et al. (2018) are introduced into the design of control protocols. Compared with the node-based event-triggered mechanism, the advantages of the edge-based event-triggered mechanism are reflected in: intermittent asynchronous communication, saving system communication resources, simple construction of the triggering function, and so on. And the disadvantages are reflected in: the significant increase in the number of judging triggering conditions, and the significant increase in the number of the designed event-triggered estimators. To summarize, the kind of triggering mechanism can be chosen according to the resources equipped by the agent itself.

1.3 Basic Knowledge

For the convenience of description, this chapter provides explanations of commonly used symbols, algebraic graph theory-related knowledge, and some necessary lemmas.

1.3.1 Symbol Description

The symbols $R^{n\times m}$, R^n, R denote $n \times m$, $n \times 1$ dimensional Euclidean space, and the set of real numbers, $\mathbb{N}$ denotes a set of non-negative integers. For given matrix A, $A > 0(\geq 0)$, $A < 0$ (≤ 0) denote the positive definite matrix (positive semi-definite matrix) and negative definite matrix (negative semi-definite matrix), $\lambda_{min}(A)$ and $\lambda_{max}(A)$ denote the minimum and maximum eigenvalues with symmetric matrix A, rank$[A]$ denotes the rank of matrix A. diag$\{A_1, \ldots, A_N\}$ denotes block diagonal matrix with $A_i \in R^{n_i \times m_i}$, $i = 1, \ldots, N$. For given $x_1 \in R^{n_1}, x_2 \in R^{n_2}, \ldots, x_N \in R^{n_N}$, $\text{col}(x_1, \ldots, x_N) \triangleq [x_1^T, \ldots, x_N^T]^T$ denotes $n_1 + n_2 + \cdots + n_N$dimensional column vector. Define convex hull $Co\{x\} = \{\sum_{i=1}^{N} o_i x_i | o_i \in R,\ o_i > 0, \sum_{i=1}^{N} o_i = 1\}$, where $x = col(x_1, \ldots, x_n)$ is the column vector. $\|\cdot\|$ denotes 2 norm for the matrix or the vector, $\otimes$ denotes the Kronecker Product, $I_N =$ diag$\{1, \ldots, N\}$, $1_N = \text{col}(1, \ldots, N)$. Suppose that $a_1, \ldots, a_N$ are monotonically increasing normal numbers, $I[a_1, \ldots, a_N] = \{a_1, a_2, \ldots, a_N\}$.

1.3.2 Topology Graph Theory

Consider a communication topology graph $\mathcal{G} = (\mathcal{V}, \mathcal{E})$ composed of N agents, where $\mathcal{V} = \{1, 2, \ldots, N\}$ and $\mathcal{E} \subseteq \mathcal{V} \times \mathcal{V}$ is the edge set. $(i, j) \in \mathcal{E}$ denotes that agent i can receive the information from agent j and $a_{ij} = 1$, $a_{ij} = 0$ otherwise. Moreover, if $(i, j) \in \mathcal{E}$, we call agent j a neighbor of agent i. $\mathcal{N}_i = \{j \in \mathcal{V}, (i, j) \in \mathcal{E}\}$ denotes neighbor sets of agent i. $\mathcal{G}$ is connected if there is a path between any two nodes in the graph $\mathcal{G}$. The adjacency matrix $\mathcal{A} = [a_{ij}]_{N\times N}$ is defined as

$$a_{ij} = \begin{cases} 1, & (i, j) \in \mathcal{E} \\ 0, & (i, j) \notin \mathcal{E} \\ 0, & i = j. \end{cases}$$

The Laplace matrix $\mathcal{L} = [l_{ij}]_{N\times N}$ is defined as

$$l_{ij} = \begin{cases} \sum_{j\in \mathcal{N}_i} a_{ij}, & i = j \\ -a_{ij}, & i \neq j. \end{cases}$$

The Laplace matrix $\mathcal{L}$ has a simple zero eigenvalue with other eigenvalues being strictly positive, that is, $0 = \lambda_1 < \lambda_2 \leq \cdots \leq \lambda_N$ if and only if topology graph $\mathcal{G}$ is connected.

Consider a communication topology graph $\bar{\mathcal{G}} = \{\bar{\mathcal{V}}, \bar{\mathcal{E}}, \bar{\mathcal{A}}\}$ composed of N follower and one leader, where $\bar{\mathcal{V}} = \mathcal{V} \cup \{0\}$ denotes the node set, $\bar{\mathcal{E}} \subseteq \bar{\mathcal{V}} \times \bar{\mathcal{V}}$ denotes the edge set, node 0 denotes the leader. Define the pinning matrix as $\mathcal{A}_0 = \text{diag}\{a_{10}, \ldots, a_{N0}\}$, where $a_{i0} > 0$ denotes the follower i can receive the leader's information, else $a_{i0} = 0$. Define the leader-follower matrix as $\mathcal{H} = \mathcal{A}_0 + \mathcal{L}$.

Consider a communication topology graph $\tilde{\mathcal{G}} = \{\tilde{\mathcal{V}}, \tilde{\mathcal{E}}, \tilde{\mathcal{A}}\}$ composed of M followers and N leaders, where $\tilde{\mathcal{V}} = \{1, 2, \ldots, N + M\}$, $\tilde{\mathcal{E}} \subseteq \tilde{\mathcal{V}} \times \tilde{\mathcal{V}}$ denote the node set and edge set, respectively, $\tilde{\mathcal{A}} = [a_{ij}]_{(N+M)\times(N+M)}$ denotes adjacency matrix. Define the Laplace matrix as $\tilde{\mathcal{L}} = [\tilde{l}_{ij}]_{(N+M)\times(N+M)}$, where $\tilde{l}_{ij} = \begin{cases} \sum_{j=1, j\neq i}^{N+M} a_{ij} & i = j, \\ -a_{ij} & i \neq j. \end{cases}$ $\mathfrak{L} = \{1, \ldots, N\}$ denotes the node set composed of the leaders, $\mathfrak{F} = \{N + 1, \ldots, N + M\}$ denotes the node set composed of the followers.

1.3.3 Relative Lemmas

Lemma 1.1 (Li and Duan 2014) *For the given topology $\mathcal{G}$, 0 is the eigenvalue of the Laplace matrix $\mathcal{L}$, and the corresponding eigenvector is 1_N and its all non-zero eigenvalues have positive real parts. If topology $\mathcal{G}$ is directed, 0 is a single eigenvalue of $\mathcal{L}$ if and only if $\mathcal{G}$ contains a directed spanning tree. If topology $\mathcal{G}$ is undirected, 0 is a single eigenvalue of $\mathcal{L}$ if and only if $\mathcal{G}$ is connected. The minimum non-zero eigenvalue of the Laplace matrix $\mathcal{L}$ satisfies $\lambda_2(\mathcal{L}) = \min_{x\neq 0,\ 1_N^T x=0}(x^T \mathcal{L} x / x^T x)$.*

Lemma 1.2 (Su and Huang 2012a) *If the communication topology $\bar{\mathcal{G}}$ contains a directed spanning tree rooted at node 0, then all eigenvalues of $\mathcal{H}$ have positive real parts.*

Lemma 1.3 (Li et al. 2015) *Assuming that the communication topology $\tilde{\mathcal{G}}$ between M followers and N leaders is directed. For each follower, there is at least one directed path from the leader to it, and the leaders do not communicate with each other. In this case, the Laplace matrix can be divided into*

$$\mathcal{L} = \begin{bmatrix} \mathcal{L}_1 & \mathcal{L}_2 \\ 0_{N\times M} & 0_{N\times N} \end{bmatrix},$$

where $\mathcal{L}_1 \in R^{M\times M}$, $\mathcal{L}_2 \in R^{M\times N}$, and all eigenvalues of $\mathcal{L}_1$ are distributed in the right half plane, and the sum of the elements in each row for $\mathcal{L}_1$ is equal. All rows of the matrix $-\mathcal{L}_1^{-1}\mathcal{L}_2$ are identical and equal to $-\mathcal{L}_1^{-1}\mathcal{L}_2 = 1_M \otimes ([a_{M+1}, \ldots, a_{M+N}]/\sum_{j\in I[M+1,M+N]a_j})$, where $a_j (j \in I[M + 1, M + N])$ are the positive constants.

Suppose that the multiagent system is composed of 1 virtual leader, N (informed/uninformed) formation leaders, and M followers, where the formation leaders mark as $1, \ldots, O$ ($O < N$) (directly obtain virtual leader information), the remaining formation leaders marked as $O + 1, \ldots, N$. Note that virtual leader will not receive information from any agents. Then, the Laplace matrix can be divided into

$$\tilde{\mathcal{L}} = \begin{bmatrix} 0_{O\times O} & 0_{O\times(N-O)} & \\ \tilde{\mathcal{L}}_{2(N-O)\times O} & \tilde{\mathcal{L}}_{1(N-O)\times(N-O)} & 0_{N\times M} \\ \tilde{\mathcal{L}}_{3M\times N} & & \tilde{\mathcal{L}}_{4M\times M} \end{bmatrix}.$$

Lemma 1.4 (Wen et al. 2016) *For given directed topology $\tilde{\mathcal{G}}$, there exists at least one informed formation leader with a path to any uninformed formation leaders, and at least one formation leader with a path to any follower. Then, all eigenvalues of matrix $\tilde{\mathcal{L}}_1$ and $\tilde{\mathcal{L}}_4$ have positive real parts, $\tilde{\mathcal{L}}_1$ and $\tilde{\mathcal{L}}_4$ are nonsingular, each element of $-\tilde{\mathcal{L}}_4^{-1}\tilde{\mathcal{L}}_3$ is non-negative, and the row sum is 1. In addition, there exists positive definite diagonal matrix $\bar{\Xi} = diag\{\bar{\nu}_{O+1}, \ldots, \bar{\nu}_N\}$ such that $\bar{\Xi}\tilde{\mathcal{L}}_1 + \tilde{\mathcal{L}}_1^T\bar{\Xi}$ is positive definite matrix and satisfies $\bar{\nu} = col(\bar{\nu}_{O+1}, \ldots, \bar{\nu}_N) = (\tilde{\mathcal{L}}_1^T)^{-1}1_{N-O}$, there exists positive definite matrix $\tilde{\Xi} = diag\{\tilde{\nu}_{N+1}, \ldots, \tilde{\nu}_{N+M}\}$ such that $\tilde{\Xi}\tilde{\mathcal{L}}_4 + \tilde{\mathcal{L}}_4^T\tilde{\Xi}$ is positive matrix and satisfy $\tilde{\nu} = col(\tilde{\nu}_{N+1}, \ldots, \tilde{\nu}_{N+M}) = (\tilde{\mathcal{L}}_4^T)^{-1}1_M$.*

Lemma 1.5 (Barbalat lemma (Khalil and Grizzle 2002)) *Assuming that $\gamma(\cdot) : [0, \infty) \to R$ is a uniformly continuous function, if*

$$\lim_{t\to\infty}\int_0^t \gamma(\tau)\mathrm{d}\tau$$

exists and limited, then $\lim_{t\to\infty}\gamma(t) = 0$.

Lemma 1.6 (Young's inequality (Qian et al. 2020)) *If there exists two positive real numbers p and q satisfying $1/p + 1/q = 1$, then one obtains that $ab \le a^p/p + b^q/q$.*

Lemma 1.7 (Deng et al. 2022) *For given continuous function $V(t)$, if the following conditions hold:*

(1) $\dot{V}(t) \le -\gamma_1 V(t) + f_1(t)$, where $t \in T_s(0, \infty)$,

(2) $\dot{V}(t) \le \gamma_2 V(t) + f_2(t)$, where $t \in T_a(0, \infty)$,

γ_1, γ_2 are positive constants, $t \in T_s(0, \infty)$, $t \in T_a(0, \infty)$ denote the unattacked and attacked intervals. Then, $V(t)$ satisfies that

$$V(t) \le \int_{T_s(0,t)} f_1(\tau)e^{\gamma_2|T_a(\tau,t)|-\gamma_1|T_s(\tau,t)|}d\tau + \int_{T_a(0,t)} f_2(\tau)e^{\gamma_2|T_a(\tau,t)|-\gamma_1|T_s(\tau,t)|}d\tau + e^{-\gamma_1|T_s(0,t)|}e^{\gamma_2|T_a(0,t)|}V(0),\ \forall t \ge 0,$$

where $\int_{\mathcal{T}} hd\tau = \int_{\mathcal{T}} h(s)d\tau(s)$ denotes the Lebesgue integral, and h is a measurable function defined on the compact set $\mathcal{T}$.

Lemma 1.8 (Schur complement lemma Boyd et al. (1994)) *For given symmetric matrix* $A = \begin{bmatrix} A_{11} & A_{12} \\ A_{12}^T & A_{22} \end{bmatrix}$, *where* A_{11}, A_{22} *is square matrix. Then, the following conditions are equivalent:*

(1) $A < 0$,

(2) $A_{11} < 0,\ A_{22} - A_{12}^T A_{11}^{-1} A_{12} < 0$,

(3) $A_{22} < 0,\ A_{11} - A_{12} A_{22}^{-1} A_{12}^T < 0$.

1.4 Publication Outline

For the multiagent system with limited bandwidth and large-scale network, its distributed cooperative control problem is studied by using adaptive event-triggered strategy. According to the way of communication network, this paper is divided into two categories: directed and undirected graphs. Chapter 2 investigates the homogeneous linear multiagent system consensus problem under the case of communication link security and communication link faults, respectively. Chapter 3 considers the consensus problem for nonlinear multiagent systems with Lipschitz dynamics and one-sided Lipschitz dynamics, respectively. The first two chapters consider the consensus problem of multiagent systems in the ideal environment. For heterogeneous linear multiagent systems where disturbances, DoS attacks, and actuator faults coexist, Chap. 4 investigates the output regulation problem. It is worth noting that the first three chapters consider multiagent systems that include only one leader system, or no leader system. In fact, some scenarios can require more than one leader in order to work together on the cooperative control problem. For this reason, Chaps. 5 through 7 study the containment problem, the formation tracking problem, and the formation containment problem, respectively, for multiagent systems containing multiple leaders. The specific research content of each chapter is as follows:

This chapter gives the background and significance of multiagent systems; introduces the research progress of the consensus problem, output regulation problem, containment problem, formation problem, and event-triggered mechanism of multiagent systems; and gives the symbol description of this book, the knowledge related to the algebraic graph theory, and some commonly used lemmas.

Chapter 2 considers the consensus problem of homogeneous linear multiagent systems under communication link security and fault scenarios, respectively. For the communication link security scenario, the adaptive event-triggered consensus control scheme is adopted to achieve the consensus problem of multiagent system under switching topologies. The control strategy requires that the dimension of the information exchanged between the agents is the same as the dimension of the output information. Therefore, the proposed method requires a light communication load especially for the case of single output with state dimension greater than 2. In addition, in order to realize the intermittent communication between the agents, the model-based event-triggered mechanism is designed to reduce the consumption of system communication resources. For the case of communication link fault, the distributed observer based on the adaptive event-triggered mechanism is proposed to compensate for the impact of communication link faults on the system, based on which a control protocol is designed to solve the consensus problem of multiagent systems with communication link faults. In addition, the feasibility of the proposed method is theoretically verified through Lyapunov stability theory. At the end of this chapter, two numerical simulations are given to verify the effectiveness of the two proposed control strategies to solve the consensus problem of the multiagent system, respectively.

Chapter 3 investigates the consensus problem of nonlinear multiagent systems with Lipschitz dynamics and one-sided Lipschitz dynamics, respectively. For the case with Lipschitz dynamics, in order to enable intermittent transmission of information from the leader to the informed followers and intermittent asynchronous communication among the agents, we design node-based event-triggered mechanisms and edge-based event-triggered mechanisms for the leader and the edges in the communication network, respectively. The designed triggering mechanisms effectively save the communication resources of the multiagent systems. To solve the consensus problem, we propose a static edge-based event-triggered consensus protocol and an edge-based event-triggered consensus protocol, respectively, and theoretically demonstrate that the designed control strategies are both able to ensure that the consensus error eventually converges to zero. For the case with one-sided Lipschitz dynamics, we propose the node-based event-triggered leader-follower consensus protocol and the leaderless consensus protocol, which solve the leader-follower consensus problem and the leaderless consensus problem, respectively, for nonlinear multiagent systems with one-sided Lipschitz dynamics. At the end of this chapter, the proposed edge-based event-triggered consensus protocol is applied to the DC motor-driven single-link manipulator model with rotary joints to validate that the proposed control strategy can solve the consensus problem for multiagent systems with Lipschitz dynamics. Similarly, the proposed node-based event-triggered leaderless consensus protocol is applied to a numerical simulation of a nonlinear multiagent systems with one-Lipschitz dynamics to validate the effectiveness and practicality of the proposed method.

Chapter 4 investigates the output regulation problem for heterogeneous linear multiagent systems where DoS attacks and actuator faults coexist. This chapter considers the more general case where only subsystems directly connected to the leader have access to the leader's matrix and state information. In order to address the impact arising from DoS attacks, this chapter designs distributed edge-based event-triggered observers for the followers to estimate the leader's matrix and state. For the case where the system matrix of the leader is unknown to some of the agents, a solution computation method for solving the regulator equation online is given. Considering the possible faults of the actuators, an adaptive event-triggered fault-tolerant control scheme is proposed to compensate for the unfavorable effects of actuator faults. Finally, it is shown that the proposed event-triggered output feedback fault-tolerant controller can effectively compensate the adverse effect of actuator faults on the system stability and realize the intermittent controller update through Lyapunov stability theory.

Chapter 5 studies the containment problem for homogeneous linear multiagent systems, where the leader has zero and non-zero input cases are considered separately. Firstly, in order to simultaneously realize intermittent communication and intermittent controller update of the agents, instead of designing a double event-triggered mechanism, based on the relative information between the agents, a new distributed adaptive event-triggered consensus protocol is proposed. The design of the event-triggered mechanism also utilizes relative information, a drawback is that it requires continuous measurement of relative information. In order to avoid this drawback, a new method that only needs to measure the relative information at the triggering moment is given. Secondly, it is theoretically demonstrated that the proposed method can solve the containment problem of homogeneous linear multiagent systems. Thirdly, the case of a zero input leader is generalized to the case of a non-zero input leader. The analysis shows that the proposed method can solve the containment problem for homogeneous linear multiagent systems with non-zero input leaders. Finally, the effectiveness of the proposed method is verified by applying it to an inverted pendulum model, and comparative simulations are given to further demonstrate the advantages of the method in this chapter.

In Chap. 6, the output formation tracking problem for heterogeneous linear multiagent systems with disturbances is investigated through adaptive dynamic event-triggered control. The goal of this chapter is to make the trajectories of all the followers form a predefined formation queue and simultaneously track on the convex combination consisting of the leader's output. To estimate the leaders' convex combination, the distributed compensators are designed. To reduce the loss of communication resources of the multiagent system, the dynamic event-triggered mechanism is designed, in which internal variables are introduced and thus dynamically adjust the threshold value of each follower. Note that the time-varying thresholds proposed in this chapter ensure a larger adjacent triggering intervals compared to the traditional static thresholds. Then, the output formation tracking protocol is designed based on the proposed compensator, the compensation input, and the time-varying formation signal, where the compensation input provides a degree of freedom for the feasible formation set. Finally, two examples are presented to validate the feasibility and effectiveness of the methods proposed in this chapter.

Chapter 7 investigates the formation containment problem for heterogeneous linear multiagent systems, where the leader dynamics are heterogeneous and classified into three categories, namely, the virtual leader, the informed formation leaders, and the uninformed formation leaders. And observers are designed to estimate the states of virtual leader for informed and uninformed formation leaders, respectively. Secondly, the formation control protocol is designed based on the proposed observer to enable the formation leaders to form a desired formation. Again, designing distributed containment controllers for the followers finally solve the formation containment problem and analyze the effectiveness of the proposed method. Finally, the simulation example verifies that the proposed method in this chapter can effectively solve the formation containment problem for heterogeneous linear multiagent systems.

Chapter 8 concludes the main result of this book and gives some problems and possible future directions of this research in terms of theoretical studies and practical applications. At the end of the chapter, we discuss future research work.

References

Adaldo A, Alderisio F, Liuzza D, Shi G, Dimarogonas DV, diBernardo M, Johansson KH (2015) Event-triggered pinning control of switching networks. IEEE Trans Control Netw Syst 2(2):204–213

Boyd S, Ghaoui LEI, Feron E, Balakrishnan V (1994) Linear matrix inequalities in system and control theory. SIAM, Philadelphia, PA

Cai H, Lewis F, Hu G, Huang J (2017) The adaptive distributed observer approach to the cooperative output regulation of linear multi-agent systems. Automatica 75:299–305

Cao Y, Stuart D, Ren W, Meng Z (2011) Distributed containment control for multiple autonomous vehicles with double-integrator dynamics: algorithms and experiments. IEEE Trans Control Syst Technol 19(4):929–938

Cao Y, Ren W, Egerstedt M (2012) Distributed containment control with multiple stationary or dynamic leaders in fixed and switching directed networks. Automatica 48(8):1586–1597

Cheng B, Li Z (2018) Fully distributed event-triggered protocols for linear multiagent networks. IEEE Trans Autom Control 64(4):1655–1662

Cheng B, Li Z (2019a) Fully distributed event-triggered protocols for linear multiagent networks. IEEE Trans Autom Control 64(4):1655–1662

Cheng B, Li Z (2019b) Coordinated tracking control with asynchronous edge-based event-triggered communications. IEEE Trans Autom Control 64(10):4321–4328

Cheng B, Wu Z, Li Z (2021) Distributed edge-based event-triggered formation control. IEEE Trans Cybern 51(3):1241–1252

DeLellis P, di Bernardo M, Gorochowski TE, Russo G (2010) Synchronization and control of complex networks via contraction, adaptation and evolution. IEEE Circuits Syst Mag 10(3):64–82

Deng C, Zhang D, Feng G (2022) Resilient practical cooperative output regulation for mass with unknown switching exosystem dynamics under dos attacks. Automatica 139:110172

Dimarogonas D, Tsiotras P, Kyriakopoulos K (2009) Leader-follower cooperative attitude control of multiple rigid bodies. Syst Control Lett 58(6):429–435

Dimarogonas DV, Frazzoli E, Johansson KH (2012) Distributed event-triggered control for multi-agent systems. IEEE Trans Autom Control 57(5):1291–1297

Dimarogonas G, Egerstedt M, Kyriakopoulos K (2006) A leader-based containment control strategy for multiple unicycles. In: Proceedings of the 45th IEEE conference on decision and control, San Diego, CA, USA, pp 5968–5973

Ding L, Han Q-L, Ge X, Zhang X-M (2018) An overview of recent advances in event-triggered consensus of multiagent systems. IEEE Trans Cybern 48(4):1110–1123

Dong X, Hu G (2017) Time-varying formation tracking for linear multiagent systems with multiple leaders. IEEE Trans Autom Control 62(7):3658–3664

Fax J, Murray R (2004) Information flow and cooperative control of vehicle formations. IEEE Trans Autom Control 49(9):1465–1476

Garcia E, Antsaklis PJ (2013) Model-based event-triggered control for systems with quantization and time-varying network delays. IEEE Trans Autom Control 58(2):422–434

Garcia E, Cao Y, Casbeer DW (2017) Periodic event-triggered synchronization of linear multi-agent systems with communication delays. IEEE Trans Autom Control 62(1):366–371

Hu W, Liu L, Feng G (2016) Consensus of linear multi-agent systems by distributed event-triggered strategy. IEEE Trans Cybern 46(1):148–157

Hua Y, Dong X, Wang J, Li Q, Zhang R (2018) Time-varying output formation tracking of heterogeneous linear multi-agent systems with multiple leaders and switching topologies. J Franklin Inst 356(1):539–560

Hua Y, Dong X, Hu G, Li Q, Ren Z (2019) Distributed time-varying output formation tracking for heterogeneous linear multiagent systems with a nonautonomous leader of unknown input. IEEE Trans Autom Control 64(10):4292–4299

Jiang W, Wen G, Peng Z, Huang T, Rahmani A (2019) Fully distributed formation-containment control of heterogeneous linear multiagent systems. IEEE Trans Autom Control 64(9):3889–3896

Jiang D, Wen G, Peng Z, Huang T, Rahmani A (2022) Fully distributed dual-terminal event-triggered bipartite output containment control of heterogeneous systems under actuator faults. IEEE Trans Syst Man Cybern Syst 52(9):5518–5531

Khalil HK, Grizzle JW (2002) Nonlinear systems, vol 3. Prentice hall Upper Saddle River, NJ

Li Y (2004) Multi-agent systems and their applications in predictive control and intelligent transportation systems. East China University of Science and Technology Press, Shanghai

Li Z, Duan Z (2014) Cooperative control of multi-agent systems: a consensus region approach. CRC Press, Boca Raton, FL, USA

Li Z, Ren W, Liu X, Fu M (2012) Consensus of multi-agent systems with general linear and Lipschitz nonlinear dynamics using distributed adaptive protocols. IEEE Trans Autom Control 58(7):1786–1791

Li Z, Ren W, Liu X, Fu M (2013) Distributed containment control of multi-agent systems with general linear dynamics in the presence of multiple leaders. Int J Robust Nonlinear Control 23:534–547

Li Z, Duan Z, Ren W, Feng G (2015) Containment control of linear multi-agent systems with multiple leaders of bounded inputs using distributed continuous controllers. Int J Robust Nonlinear Control 25:2101–2121

Li Z, Chen M, Ding Z (2016) Distributed adaptive controllers for cooperative output regulation of heterogeneous agents over directed graphs. Automatica 68:179–183

Li R, Zhang L, Wang J (2018) A control method of unmanned car following under time-varying relative distance and angle. Acta Automatica Sinica 11:2031–2040

Li X, Tang Y, Karimi H (2020) Consensus of multi-agent systems via fully distributed event-triggered control. Automatica 116:1–8

Liu A, Qin D (2021) Distributed predictive control of multiple mobile robots based on virtual structure method. Control Decis 5:1273–1280

Liu K, Ji Z, Ren W (2017) Necessary and sufficient conditions for consensus of second-order multiagent systems under directed topologies without global gain dependency. IEEE Trans Cybern 47(8):2089–2098

Liu C-L, Shan L, Chen Y-Y, Zhang Y (2018) Average-consensus filter of first-order multi-agent systems with disturbances. IEEE Trans Circuits Syst II: Express Briefs 65(11):1763–1767

Lv Y, Li Z, Duan Z, Chen J (2016) Distributed adaptive output feedback consensus protocols for linear systems on directed graphs with a leader of bounded input. Automatica 74:308–314

Ma Q, Lewis F, Xu S (2015) Cooperative containment of discrete-time linear multi-agent systems. Int J Roboust Nonlinear Control 25:1007–1018

Meng Z, Ren W, You Z (2010) Distributed finite-time attitude containment control for multiple rigid bodies. Automatica 46(12):2092–2099

Meng X, Xie L, Soh YC (2018) Event-triggered output regulation of heterogeneous multiagent networks. IEEE Trans Autom Control 63(12):4429–4434

Ni W, Cheng D (2010) Leader-following consensus of multi-agent systems under fixed and switching topologies. Syst Control Lett 59:209–217

Olfati-Saber R, Murray R (2004) Consensus problems in networks of agents with switching topology and time-delays. IEEE Trans Autom Control 49(9):1520–1533

Qian Y-Y, Liu L, Feng G (2019) Output consensus of heterogeneous linear multi-agent systems with adaptive event-triggered control. IEEE Trans Autom Control 64(6):2606–2613

Qian Y-Y, Liu L, Feng G (2020) Distributed event-triggered adaptive control for consensus of linear multi-agent systems with external disturbances. IEEE Trans Cybern 50(5):2197–2208

Qian Y-Y, Liu L, Feng G (2021) Cooperative output regulation of linear multiagent systems: an event-triggered adaptive distributed observer approach. IEEE Trans Autom Control 66(2):833–840

Rehan M, Jameel A, Ahn CK (2018) Distributed consensus control of one-sided Lipschitz nonlinear multiagent systems. IEEE Trans Syst Man Cybern Syst 48(8):1297–1308

Ren W, Beard R (2015) Consensus seeking in multiagent systems under dynamically changing interaction topologies. IEEE Trans on Autom Control 50(5):655–661

Ren W, Moore K, Chen Y (2006) High-order consensus algorithms in cooperative vehicle systems. In: Proceedings of the 2006 IEEE international consensus on IEEE, pp 457–462

Reynolds G (1987) Flocks, birds and schools: a distributed behavioral model. Comput Graphics 21(1):25–34

Rogers A, Corkill DD, Jennings NR (2009) Agent technologies for sensor networks. IEEE Intell Syst 24(2):13–17

Seyboth GS, Dimarogonas DV, Johansson KH (2013) Event-based broadcasting for multi-agent average consensus. Automatica 49(1):245–252

Shi L, Shao J, Zheng W, Huang T (2017) Asynchronous containment control for discrete-time second-order multi-agent systems with time-varying delays. J Franklin Inst 354(18):8552–8569

Song Y, Yang F (2012) Research on configuration control of multi-agent system based on behavior method. Control Eng 4:687–690

Su Y, Huang J (2012a) Cooperative output regulation of linear multi-agent systems. IEEE Trans Automatica Control 57(4):1062–1066

Su Y, Huang J (2012b) Cooperative output regulation with application to multi-agent consensus under switching network. IEEE Trans Syst Man Cybern. Part B (Cybern) 42(3):864–875

Tabuada P (2007) Event-triggered real-time scheduling of stabilizing control tasks. IEEE Trans Autom Control 52(9):1680–1685

Vicsek G, Czirok A, Benjacob E (1995) Novel type of phase-transition in a systme of self-driven particles. Phys Rev Lett 75(6):1226–1229

Wang R, Dong X, Li Q, Ren Z (2019) Distributed time-varying output formation control for general linear multiagent systems with directed topology. IEEE Trans Control Netw Syst 6(2):609–620

Wen G, Zhao Y, Duan Z, Yu W, Chen G (2016) Containment of higher-order multi-leader multi-agent systems: a dynamic output approach. IEEE Trans Autom Control 61(4):1135–1140

Wu Y, Cao X, Zheng P, Zeng Z (2012) Variable structure-based decentralized relative attitude-coordinated control for satellite formation. IEEE Aerosp Electron Syst Mag 27(12):18–25

Xiong G, Wen G, Peng Z, Huang T (2022) Pull-based event-triggered containment control for multiagent systems with active leaders via aperiodic sampled-data transmission. IEEE Trans Syst Man Cybern Syst 52(1):214–225

Yang D, Ren W, Liu X, Chen W (2016) Decentralized event-triggered consensus for linear multi-agent systems under general directed graphs. Automatica 69:242–249

Yang R, Zhang H, Feng G, Yan H (2018) Distributed event-triggered adaptive control for cooperative output regulation of heterogeneous multiagent systems under switching topology. IEEE Trans Neural Netw Learn Syst 29(9):4347–4358

Yu H, Antsaklis PJ (2014) Output synchronization of networked passive systems with event-driven communication. IEEE Trans Autom Control 59(3):750–756

Yu W, Chen G, Ren W, Kurths J, Zheng WX (2011a) Distributed higher order consensus protocols in multiagent dynamical systems. IEEE Trans Circuits Syst I: Regular Papers 58(8):1924–1932

Yu W, Zheng WX, Chen G, Ren W, Cao J (2011b) Second-order consensus in multi-agent dynamical systems with sampled position data. Automatica 47(7):1496–1503

Zhang H, Feng G, Yan H, Chen Q (2014) Observer-based output feedback event-triggered control for consensus of multi-agent systems. IEEE Trans Ind Electron 61(9):4885–4894

Zhang J, Zhang H, Sun S, Cai Y (2022) Adaptive time-varying formation tracking control for multiagent systems with nonzero leader input by intermittent communications. IEEE Trans Cybern 1–10. https://doi.org/10.1109/TCYB.2022.3165212

Zhou S, Cheng Y, Lei X, Duan H (2022) Multi-agent few-shot meta reinforcement learning for trajectory design and channel selection in uav-assisted networks. China Commun 19(4):166–176

Zhu W, Jiang Z-P, Feng G (2014) Event-based consensus of multi-agent systems with general linear models. Automatica 50(2):552–558

Zuo S, Song Y, Lewis FL, Davoudi A (2019) Time-varying output formation containment of general linear homogeneous and heterogeneous multiagent systems. IEEE Trans Control Netw Syst 6(2):537–548

Chapter 2
Consensus Control of Linear Multiagent Systems Under Switching Networks and Communication Link Faults

2.1 Introduction

Consensus, as one of the most fundamental issues in collaborative control, aims to design appropriate distributed control schemes through mutual communication between agents, so that a group of agents can achieve consensus in state or output. So far, there have been many reports on the study of consensus in multiagent systems (Ni and Cheng 2010; Wen et al. 2014; Zhang et al. 2021; Li et al. 2010a). For example, the authors in Ni and Cheng (2010) study the leader-follower consensus problem in high-order multiagent systems for fixed and switched communication topologies. The authors in Wen et al. (2014) study the H_∞ consensus problem of multiagent systems under directed graphs. However, the methods in these results not only require continuous communication but also rely on the global information of the communication topology, such as the eigenvalues of the Laplacian matrix.

Note that there are already many results on event-triggered control for multiagent systems in Adaldo et al. (2015), Liuzza et al. (2013), Yi et al. (2017), Zhang et al. (2019), Meng and Chen (2013), and Liu et al. (2020), where agents avoid continuous communication and/or controller updates. Only when the measurement error exceeds the allowable range, the agent will send information to all its neighbors and update the controller. For example, the authors in Adaldo et al. (2015) proposed a cloud-based event-triggered control algorithm to solve the practical consensus problem of multi-agent systems with integrator dynamics. The authors in Liuzza et al. (2013) proposed a model-based event-triggered control scheme to solve the practical consensus problem of nonlinear multi-agent systems. However, these literature share a common drawback that the construction of control protocols and triggering mechanisms relied on the minimum and maximum non-negative eigenvalues of the Laplacian matrix. For each agent, it is necessary to traverse the entire communication topology to calculate the eigenvalues of the Laplacian matrix, which is unrealistic for multi-agent systems with large-scale networks. In order to avoid dependence on

J. Zhang et al., *Adaptive Collaborative Control of Multiagent Systems by Event-Triggered Mechanisms*, Intelligent Control and Learning Systems 14,
https://doi.org/10.1007/978-981-97-7866-9_2

the global information of the communication topology, an adaptive control algorithm was introduced to solve the collaborative control problem of multiagent systems.

Pioneer work on consensus results is obtained under secure communication topologies. However, topology switching and link faults may exist in multiagent system, which is a more common situation. Therefore, studying the consensus control problem of multiagent system with switching topologies and communication link faults is important and meaningful.

2.2 Problem Formulation

Consider a multiagent system composed of N agents, the ith subsystem is modeled as

$$\begin{cases} \dot{x}_i(t) = Ax_i(t) + Bu_i(t) \\ y_i(t) = Cx_i(t),\ i \in I[1, N] \end{cases} \tag{2.1}$$

in which $x_i(t) \in R^n$, $u_i(t) \in R^p$, $y_i(t) \in R^q$ denote the system state, input, output, and $A,\ B,\ C$ are known system matrices.

The aim of the chapter is to design the adaptive event-triggered control protocol under switching topologies or communication link faults such that the following equalities satisfy

$$\lim_{t\to\infty} \|x_i(t) - x_j(t)\| = 0,\ \forall\, i, j \in I[1, N] \tag{2.2}$$

or

$$\lim_{t\to\infty} \|x_i(t) - x_0(t)\| = 0,\ \forall\, i \in I[1, N], \tag{2.3}$$

the Zeno behavior is excluded under designed triggering mechanisms, where x_0 in Eq. (2.3) denotes the leader state satisfying $\dot{x}_0(t) = Ax_0(t)$.

The leaderless consensus is reached if Eq. (2.2) holds. The leader-follower consensus is reached if Eq. (2.3) holds. On the premise of not causing ambiguity, time t is omitted later in this chapter.

2.3 Consensus Under Switching Topologies

In this section, we provide the design of controllers to achieve leaderless consensus and leader-follower consensus in multi-agent systems under switching topologies. In order to avoid redundant calculation of global information for switching topologies, event-triggered control protocols based on adaptive observers are constructed separately.

2.3.1 Leaderless Consensus

To achieve the leaderless consensus, the following assumptions are given.

Assumption 2.1 The matrix pair (A, B) is stabilizable and (A, C) is detectable.

Assumption 2.2 The communication graph $\mathcal{G}^{\sigma(t)}$ is fixed and connected at switching time interval $[t_g, t_{g+1})$, $\sigma(t)$ is switching signal. $t_g(t_1 = 0)$ is the switching instant for g being finite non-negative integer. Moreover, $0 < \tau < t_{g+1} - t_g < T$, where $\tau > 0$ is the dwell time, and $T > 0$.

Then, design the following adaptive observer-based event-triggered control protocol for agent i as follows:

$$\begin{cases} u_i = G\xi_i \\ \dot{\xi}_i = (A + BG)\xi_i + EC \sum\limits_{j \in N_i} d_{ij} a_{ij}^{\sigma(t)} ((\hat{\xi}_i - \hat{\xi}_j) - (\hat{\varsigma}_i - \hat{\varsigma}_j)) \\ \dot{d}_{ij} = k_{ij} a_{ij}^{\sigma(t)} \begin{pmatrix} \hat{\varsigma}_i - \hat{\varsigma}_j \\ \hat{\xi}_i - \hat{\xi}_j \end{pmatrix}^T \mathcal{R} \begin{pmatrix} \hat{\varsigma}_i - \hat{\varsigma}_j \\ \hat{\xi}_i - \hat{\xi}_j \end{pmatrix} \\ \dot{\varsigma}_i = A\varsigma_i + Bu_i + F(C\varsigma_i - y_i) \end{cases} \tag{2.4}$$

in which ξ_i, ς_i express the states of controller and observer, respectively, d_{ij} $d_{ij}(0) = d_{ji}(0)$ expresses the adaptive coupling weight for the edge, E, F, G, $\mathcal{R}$ are the designed matrices, $k_{ij} = k_{ji}$ are given positive constants, $\hat{\xi}_i$, $\hat{\varsigma}_i$ express the estimators for ξ_i and ς_i satisfying

$$\begin{cases} \hat{\varsigma}_i = e^{A(t - t_k^i)} \varsigma_i(t_k^i), \ t \in [t_k^i, t_{k+1}^i) \\ \hat{\xi}_i = e^{A(t - t_k^i)} \xi_i(t_k^i), \ t \in [t_k^i, t_{k+1}^i) \end{cases} \tag{2.5}$$

with $t_k^i(t_0^i = 0)$ being triggering instant depending on the triggering rule as follows:

$$t_{k+1}^i = \begin{cases} inf\{t > t_{k'}^i | h_i \geq 0\} \\ t_g, \text{if graph is switching,} \end{cases} \tag{2.6}$$

and h_i is the triggering function to be constructed.

Remark 2.1 The idea of the constructing control protocol in Eq. (2.3) was inspired by Zhang et al. (2013) and Li et al. (2013). In detail, compared with Zhang et al. (2013), Eq. (2.3) can effectively avoid the repeated calculation of global information in the communication topology by introducing adaptive weights. Compared with Li et al. (2013), controller (2.3) can reduce the information transmission between neighboring agents by introducing event-triggered estimators $\hat{\xi}_i$ and $\hat{\varsigma}_i$.

Remark 2.2 In the early stage, although adaptive algorithms emerged to avoid using global knowledge of communication topology in complex network synchronization problems, the control protocols designed were centralized in Chen et al. (2007). Centralized control has lower fault tolerance because when the computer implementing centralized control fails, the entire system would be paralyzed, and economic losses will be incalculable. Subsequently, the adaptive algorithm was introduced into MAS to fully distribute the control protocol (Li et al. 2013). The fully distributed control protocol avoids knowing the entire multi-agent communication network and has strong fault tolerance.

For simplicity, the following definition is provided. Set $z_i = \mathrm{col}(\varsigma_i, \xi_i)$, $z = \mathrm{col}(z_1, \ldots, z_N)$, $\bar{z}_i = \mathrm{col}(x_i, \xi_i)$, $\bar{z} = \mathrm{col}(\bar{z}_1, \ldots, \bar{z}_N)$, $\hat{z}_i = \mathrm{col}(\hat{\varsigma}_i, \hat{\xi}_i)$, $\hat{z} = \mathrm{col}(\hat{z}_1, \ldots, \hat{z}_N)$, $\eta_i = z_i - \frac{1}{N} \sum\limits_{l \in I[1,N]} z_l$, $\eta = \mathrm{col}(\eta_1, \ldots, \eta_N)$, $\bar{\eta}_i = \bar{z}_i - \frac{1}{N} \sum\limits_{i \in I[1,N]} \bar{z}_l$ or $\bar{\eta}_i = \bar{z}_i - \sum\limits_{i \in I[1,N]} \alpha_i \bar{z}_l$ with $\sum\limits_{i \in I[1,N]} \alpha_i = 1$, $\bar{\eta} = \mathrm{col}(\bar{\eta}_1, \ldots, \bar{\eta}_N)$.

Then, the following compact form can be obtained:

$$\begin{cases} \eta = \left((I_N - \frac{1}{N}\mathbf{1} \cdot \mathbf{1}^T) \otimes I_{2n}\right) z \\ \bar{\eta} = \left((I_N - \frac{1}{N}\mathbf{1} \cdot \mathbf{1}^T) \otimes I_{2n}\right) \bar{z}. \end{cases} \tag{2.7}$$

It's easy to obtain that $\eta = 0$ if and only if $z_1 = \cdots = z_N$, and $\bar{\eta} = 0$ if and only if $\bar{z}_1 = \cdots = \bar{z}_N$, which indicates that $x_1 = \cdots = x_N$ if $\eta = 0$ and $\bar{\eta} = 0$. Consequently, $\bar{\eta}$ is called as the consensus error. Therefore, leaderless consensus is reached if $\lim_{t \to \infty} \bar{\eta} = 0$. This idea was derived from reference Lu and Chen (2006), which proposed a general framework for analyzing synchronization issues of linear coupled differential dynamics. The study of leaderless consensus has laid a theoretical foundation for further research on synchronization problems in complex networks and multiagent systems (Chen et al. 2007; Lu et al. 2015; Li et al. 2010b).

Subsequently, set the observer error as $\epsilon_i = \varsigma_i - x_i$ and calculate the derivative of ϵ_i as

$$\dot{\epsilon}_i = (A + FC)\epsilon_i, \; i \in I[1, N]. \tag{2.8}$$

Then, calculate the derivative of z_i as

$$\begin{cases} \dot{z}_i = \mathcal{F} z_i + \sum\limits_{j \in I[1,N]} d_{ij} a_{ij}^{\sigma(t)} \mathcal{M}(\hat{z}_i - \hat{z}_j) + \Xi \epsilon_i \\ \dot{d}_{ij} = k_{ij} a_{ij}^{\sigma(t)} (\hat{z}_i - \hat{z}_j)^T \mathcal{R} (\hat{z}_i - \hat{z}_j) \end{cases} \tag{2.9}$$

in which $\mathcal{F} = \begin{pmatrix} A & BG \\ 0 & A + BG \end{pmatrix}$, $\mathcal{M} = \begin{pmatrix} 0 & 0 \\ -EC & EC \end{pmatrix}$, $\Xi = \begin{pmatrix} FC \\ 0 \end{pmatrix}$ are feedback gain matrices. It follows from the definition of η_i that the derivative of η_i satisfies

$$\begin{cases} \dot{\eta}_i = \mathcal{F} \eta_i + \sum\limits_{j \in I[1,N]} d_{ij} a_{ij}^{\sigma(t)} \mathcal{M}(\hat{z}_i - \hat{z}_j) + \Xi(\epsilon_i - \frac{1}{N} \sum\limits_{l \in I[1,N]} \epsilon_l) \\ \dot{d}_{ij} = k_{ij} a_{ij}^{\sigma(t)} (\hat{z}_i - \hat{z}_j)^T \mathcal{R} (\hat{z}_i - \hat{z}_j). \end{cases} \tag{2.10}$$

Set the estimate error as $e_i = \hat{z}_i - z_i$, and design the following triggering function h_i as

$$h_i = -\nu e^{-wt} - \frac{1}{4} \sum_{j \in I[1,N]} a_{ij}^{\sigma(t)} (\hat{z}_i - \hat{z}_j)^T \mathcal{R} (\hat{z}_i - \hat{z}_j) + \sum_{j \in I[1,N]} (2\iota d_{ij} + 1) a_{ij}^{\sigma(t)} e_i^T \mathcal{R} e_i \tag{2.11}$$

in which w, ι, ν are given positive constants, $i \in 1[1, N]$.

Remark 2.3 The mode-based event-triggered mechanism (2.6) is adopted in this chapter, so the proposed method avoids continuous communication between neighbors and continuous supervision of neighbor information. Moreover, the constructed triggering function (2.11) does not need any global information.

In the following text, the construction of the feedback gain matrices and parameters in the fully distributed event-triggered control scheme (2.3) and triggering function (2.7) is presented in Algorithm 2.1.

Algorithm 2.1 Suppose that Assumption 2.1 is satisfied, execute the following four steps:

Step 1: Feedback gain matrices F, G are selected such that the matrices $A + FC$ and $A + BG$ are Hurwitz.
Step 2: Matrix $P > 0$ satisfies the following linear matrix inequality (LMI):

$$PA + A^T P - 2C^T C < 0. \tag{2.12}$$

Step 3: $\mathcal{R} = \begin{bmatrix} C^T C & -C^T C \\ -C^T C & C^T C \end{bmatrix}$ and $E = -P^{-1} C^T$ are set.
Step 4: $k_{ij}, \iota, \nu, \omega$ are selected to be positive constants.

Remark 2.4 Under the condition of Assumption 2.1, Step 1 in Algorithm 2.1 can be ensured. The relevant proof is provided in detail in Lemma 4 for Zhang et al. (2020), which is omitted here. Moreover, the solution of LMI (2.12) can be ensured under the condition of Assumption 2.1.

Lemma 2.1 *The designed matrix $\mathcal{R}$ in Algorithm 2.1 is positive semi-definite.*

Proof Set $U = \begin{bmatrix} I & 0 \\ -I & I \end{bmatrix}$, $\tilde{\mathcal{R}} = (U^{-1})^T \mathcal{R} U^{-1}$. Obviously, matrix U is nonsingular, matrices $\mathcal{R}$ and $\tilde{\mathcal{R}}$ are similar, i.e., $\mathcal{R} \sim \tilde{\mathcal{R}}$. Therefore, we can conclude that the eigenvalues of matrix $\mathcal{R}$ and matrix $\tilde{\mathcal{R}}$ are the same. Then, calculate U^{-1} and $\tilde{\mathcal{R}}$ as

$$U^{-1} = \begin{bmatrix} I & 0 \\ I & I \end{bmatrix}, \tilde{\mathcal{R}} = \begin{bmatrix} 0 & 0 \\ 0 & C^T C \end{bmatrix}.$$

Since $(C^T C)^T = C^T C$, one can obtain that matrix $C^T C$ is symmetric. Therefore, for any non-zero vector y_1, one has

$$x^T C^T C x = (Cx)^T (Cx) \geq 0.$$

Thus, matrix $C^T C$ is positive semi-definite, which shows that matrix $\tilde{\mathcal{R}}$ is positive semi-definite. Furthermore, one can easily conclude that matrix $\mathcal{R}$ is positive semi-definite. □

Theorem 2.1 *Under the condition of Assumptions 2.1–2.2, and the relative parameters designed for the distributed observer-based event-triggered control scheme in Eq.* (2.3) *and triggering function in Eq.* (2.7) *are given in Algorithm 2.1 such that matrices $A + FC$ and $A + BG$ are Hurwitz,* $\mathcal{R} = \begin{bmatrix} C^T C & -C^T C \\ -C^T C & C^T C \end{bmatrix}$, $E = -P^{-1} C^T$. *Then, the leaderless consensus is reached, and the adaptive weights converge to some finite values. Moreover, the multiagent system does not exhibit the Zeno behavior under the triggering mechanism* (2.6).

Proof Based on Algorithm 2.1 in Step 1, there exists solution W such that

$$(A + FC)^T W + W(A + FC) + \gamma I = 0$$

where the parameter $\gamma > 0$ is positive. Then, select the following Lyapunov function as follows:

$$V_1 = \sum_{i \in I[1,N]} \epsilon_i^T W \epsilon_i. \tag{2.13}$$

According to (2.8), calculate the derivative of (2.13) as

$$\begin{aligned} \dot{V}_1 &= \sum_{i \in I[1,N]} \epsilon_i^T \left(W(A + FC) + (A + FC)^T W \right) \epsilon_i \\ &= -\gamma \sum_{i \in I[1,N]} \epsilon_i^T \epsilon_i \leq 0. \end{aligned} \tag{2.14}$$

It is obvious to obtain that $\lim_{t \to \infty} \epsilon_i = 0,\ i \in I[1, N]$.

Decoupling variable ϵ_i from Eq. (2.10) yields

$$\begin{cases} \dot{\eta}_i = \mathcal{F} \eta_i + \sum_{j \in I[1,N]} d_{ij} a_{ij}^{\sigma(t)} \mathcal{M}(\hat{z}_i - \hat{z}_j) \\ \dot{d}_{ij} = k_{ij} a_{ij}^{\sigma(t)} (\hat{z}_i - \hat{z}_j)^T \mathcal{R}(\hat{z}_i - \hat{z}_j). \end{cases} \tag{2.15}$$

The stability of dynamics in Eq. (2.10) and Eq. (2.15) are equivalent.

Next, we demonstrate the stability of dynamics in Eq. (2.15). Select the Lyapunov function as

$$V_2 = \frac{1}{2} \sum_{i \in I[1,N]} \eta_i^T \mathcal{P} \eta_i + \sum_{i \in I[1,N]} \sum_{j \in I[1,N]} \frac{(d_{ij} - \beta)^2}{8k_{ij}} \tag{2.16}$$

in which β is a positive constant satisfying that $\beta = \max\{\frac{4}{\lambda_2^{min}(\mathcal{L}^{\sigma(t)})}, \frac{1}{\iota}\}$. γ is a sufficiently small positive constant, $\mathcal{P}$ is selected as $\mathcal{P} = \begin{bmatrix} \gamma \bar{P} + P & -P \\ -P & P \end{bmatrix}$, and P is given in Algorithm 2.1, $\bar{P}$ is selected such that $\bar{P}(A + BG) + (A + BG)^T \bar{P} < 0$ (it can be ensured in Algorithm 2.1). The Schur complement M of $\mathcal{P}$ satisfies that $M = \gamma \bar{P} + P - (-P)P^{-1}(-P) > 0$, which is a positive definite matrix. Based on Lemma 1.8, we can conclude that matrix $\mathcal{P}$ is positive definite.

Based on (2.15) and $\mathcal{P}\mathcal{M} = -\mathcal{R}$, calculate the derivative of V_2

$$\begin{aligned} \dot{V}_2 = & \sum_{i \in I[1,N]} \eta_i^T \mathcal{P} \mathcal{F} \eta_i - \sum_{i \in I[1,N]} \sum_{j \in I[1,N]} \eta_i^T \mathcal{R} d_{ij} a_{ij}^{\sigma(t)} (\hat{z}_i - \hat{z}_j) \\ & + \sum_{i \in I[1,N]} \sum_{j \in I[1,N]} \frac{(d_{ij} - \beta)}{4} a_{ij}^{\sigma(t)} (\hat{z}_i - \hat{z}_j)^T \mathcal{R} (\hat{z}_i - \hat{z}_j). \end{aligned} \tag{2.17}$$

Then, based on the fact that $\eta_i - \eta_j = z_i - z_j$, $e_i = \hat{z}_i - z_i$, $d_{ij} = d_{ji}$, $a_{ij}^{\sigma(t)} = a_{ji}^{\sigma(t)}$ and Lemma 2.1, we can obtain that

$$\begin{aligned} & - \sum_{i \in I[1,N]} \sum_{j \in I[1,N]} \eta_i^T \mathcal{R} d_{ij} a_{ij}^{\sigma(t)} (\hat{z}_i - \hat{z}_j) \\ = & - \frac{1}{2} \sum_{i \in I[1,N]} \sum_{j \in I[1,N]} d_{ij} a_{ij}^{\sigma(t)} (\hat{z}_i - \hat{z}_j)^T \mathcal{R} (\hat{z}_i - \hat{z}_j) \\ & + \frac{1}{2} \sum_{i \in I[1,N]} \sum_{j \in I[1,N]} d_{ij} a_{ij}^{\sigma(t)} (e_i - e_j)^T \mathcal{R} (\hat{z}_i - \hat{z}_j). \end{aligned} \tag{2.18}$$

It follows from Lemma 1.6 that

$$\begin{aligned} & \sum_{i \in I[1,N]} \sum_{j \in I[1,N]} d_{ij} a_{ij}^{\sigma(t)} (e_i - e_j)^T \mathcal{R} (\hat{z}_i - \hat{z}_j) \\ \leq & \frac{1}{2} \sum_{i \in I[1,N]} \sum_{j \in I[1,N]} d_{ij} a_{ij}^{\sigma(t)} (e_i - e_j)^T \mathcal{R} (e_i - e_j) \\ & + \frac{1}{2} \sum_{i \in I[1,N]} \sum_{j \in I[1,N]} d_{ij} a_{ij}^{\sigma(t)} (\hat{z}_i - \hat{z}_j)^T \mathcal{R} (\hat{z}_i - \hat{z}_j) \end{aligned} \tag{2.19}$$

and

$$
\begin{aligned}
&\sum_{i\in I[1,N]}\sum_{j\in I[1,N]} d_{ij}a_{ij}^{\sigma(t)}(e_i-e_j)^T\mathcal{R}(e_i-e_j)\\
=&\sum_{i\in I[1,N]}\sum_{j\in I[1,N]} d_{ij}a_{ij}^{\sigma(t)}e_i^T\mathcal{R}e_i\\
&+\sum_{i\in I[1,N]}\sum_{j\in I[1,N]} d_{ij}a_{ij}^{\sigma(t)}e_j^T\mathcal{R}e_j\\
&-2\sum_{i\in I[1,N]}\sum_{j\in I[1,N]} d_{ij}a_{ij}^{\sigma(t)}e_i^T\mathcal{R}e_j\\
\leq& 4\sum_{i\in I[1,N]}\sum_{j\in I[1,N]} d_{ij}a_{ij}^{\sigma(t)}e_i^T\mathcal{R}e_i.
\end{aligned}
\tag{2.20}
$$

Bringing (2.18)–(2.20) into (2.17), one obtains

$$
\begin{aligned}
\dot{V}_2\leq&\sum_{i\in I[1,N]}\eta_i^T\mathcal{PF}\eta_i+\sum_{i\in I[1,N]}\sum_{j\in I[1,N]} d_{ij}a_{ij}^{\sigma(t)}e_i^T\mathcal{R}e_i\\
&-\frac{\beta}{4}\sum_{i\in I[1,N]}\sum_{j\in I[1,N]} a_{ij}^{\sigma(t)}(\hat{z}_i-\hat{z}_j)^T\mathcal{R}(\hat{z}_i-\hat{z}_j)\\
=&\sum_{i\in I[1,N]}\eta_i^T\mathcal{PF}\eta_i+\sum_{i\in I[1,N]}\sum_{j\in I[1,N]} d_{ij}a_{ij}^{\sigma(t)}e_i^T\mathcal{R}e_i\\
&-\frac{\beta}{4}\sum_{i\in I[1,N]}\sum_{j\in I[1,N]} a_{ij}^{\sigma(t)}(\hat{z}_i-\hat{z}_j)^T\mathcal{R}(\hat{z}_i-\hat{z}_j).
\end{aligned}
\tag{2.21}
$$

Based on the fact that $\hat{z}_i-\hat{z}_j=e_i-e_j+z_i-z_j$ and Lemma 1.6, we have the following formula:

$$
\begin{aligned}
&-\frac{\beta}{8}\sum_{i\in I[1,N]}\sum_{j\in I[1,N]} a_{ij}^{\sigma(t)}(\hat{z}_i-\hat{z}_j)^T\mathcal{R}(\hat{z}_i-\hat{z}_j)\\
=&-\frac{\beta}{8}\sum_{i\in I[1,N]}\sum_{j\in I[1,N]} a_{ij}^{\sigma(t)}(e_i-e_j)^T\mathcal{R}(e_i-e_j)\\
&-\frac{\beta}{8}\sum_{i\in I[1,N]}\sum_{j\in I[1,N]} a_{ij}^{\sigma(t)}(\eta_i-\eta_j)^T\mathcal{R}(\eta_i-\eta_j)\\
&-\frac{\beta}{4}\sum_{i\in I[1,N]}\sum_{j\in I[1,N]} a_{ij}^{\sigma(t)}(\eta_i-\eta_j)^T\mathcal{R}(e_i-e_j)\\
\leq&-\frac{\beta}{8}\sum_{i\in I[1,N]}\sum_{j\in I[1,N]} a_{ij}^{\sigma(t)}\eta_i^T\mathcal{R}(\eta_i-\eta_j)\\
&+\frac{\beta}{2}\sum_{i\in I[1,N]}\sum_{j\in I[1,N]} a_{ij}^{\sigma(t)}e_i^T\mathcal{R}e_i.
\end{aligned}
\tag{2.22}
$$

It follows from (2.21) and (2.22) that

$$\begin{aligned}\dot{V}_2 \le & \sum_{i\in I[1,N]} \eta_i^T \mathcal{P}\mathcal{F}\eta_i - \frac{\beta}{8}\sum_{i\in I[1,N]}\sum_{j\in I[1,N]} a_{ij}^{\sigma(t)}\eta_i^T \mathcal{R}(\eta_i - \eta_j) \\ & + \frac{\beta}{2}\sum_{i\in I[1,N]}\Big\{\sum_{j\in I[1,N]}(1+\frac{2\iota}{\beta\iota}d_{ij})a_{ij}^{\sigma(t)}e_i^T \mathcal{R}e_i \\ & - \frac{1}{4}\sum_{j\in I[1,N]} a_{ij}^{\sigma(t)}(\hat{z}_i - \hat{z}_j)^T \mathcal{R}(\hat{z}_i - \hat{z}_j)\Big\} \\ \le & \sum_{i\in I[1,N]} \eta_i^T \mathcal{P}\mathcal{F}\eta_i - \frac{\beta}{8}\sum_{i\in I[1,N]}\sum_{j\in I[1,N]} a_{ij}^{\sigma(t)}\eta_i^T \mathcal{R}(\eta_i - \eta_j) \\ & + \frac{\beta}{2}\sum_{j\in I[1,N]} \nu e^{-wt}. \end{aligned} \tag{2.23}$$

Set $U = \begin{bmatrix} I & 0 \\ -I & I \end{bmatrix}$, $\tilde{\eta}_i = U\eta_i$, $\tilde{\mathcal{P}} = U^{-T}\mathcal{P}U^{-1}$, $\tilde{\mathcal{F}} = U\mathcal{F}U^{-1}$, and $\tilde{\mathcal{R}} = U^{-T}\mathcal{R}U^{-1}$. Then, one can obtain the following formula:

$$\begin{aligned}& \sum_{i\in I[1,N]} \eta_i^T \mathcal{P}\mathcal{F}\eta_i - \frac{\beta}{8}\sum_{i\in I[1,N]}\sum_{j\in I[1,N]} a_{ij}^{\sigma(t)}\eta_i^T \mathcal{R}(\eta_i - \eta_j) \\ = & \frac{1}{2}\tilde{\eta}^T\big(I_N\otimes(\tilde{\mathcal{P}}\tilde{\mathcal{F}} + \tilde{\mathcal{F}}^T\tilde{\mathcal{P}}) - \frac{\beta}{4}\mathcal{L}^{\sigma(t)}\otimes\tilde{\mathcal{R}}\big)\tilde{\eta}. \end{aligned} \tag{2.24}$$

Under the condition of Assumption 2.2, we have that $\mathcal{L}^{\sigma(t)}$ is a positive semi-definite matrix from Lemma 1.1. Therefore, there must exist a unitary matrix $J^{\sigma(t)}$ with $J^{\sigma(t)}\mathcal{L}^{\sigma(t)}(J^{\sigma(t)})^T = \{0, \lambda_2(\mathcal{L}^{\sigma(t)}), \ldots, \lambda_N(\mathcal{L}^{\sigma(t)})\} \triangleq \Omega^{\sigma(t)}$, where the following relationship is satisfied: $0 < \lambda_2(\mathcal{L}^{\sigma(t)}) \le \cdots \le \lambda_N(\mathcal{L}^{\sigma(t)})$. Set $\hat{\eta} = (J^{\sigma(t)} \otimes I_{2n})\tilde{\eta}$. Then, Eq. (2.24) satisfies

$$\begin{aligned}& \sum_{i\in I[1,N]} \eta_i^T \mathcal{P}\mathcal{F}\eta_i - \frac{\beta}{8}\sum_{i\in I[1,N]}\sum_{j\in I[1,N]} a_{ij}^{\sigma(t)}\eta_i^T \mathcal{R}(\eta_i - \eta_j) \\ \le & \frac{1}{2}\sum_{i\in I[1,N]} \hat{\eta}_i^T\Big(\tilde{\mathcal{P}}\tilde{\mathcal{F}} + \tilde{\mathcal{F}}^T\tilde{\mathcal{P}} - \frac{\beta}{4}\lambda_2^{min}(\mathcal{L}^{\sigma(t)})\tilde{\mathcal{R}}\Big)\hat{\eta}_i \triangleq W(\hat{\eta}). \end{aligned}$$

Simplification can be obtained in that

$$\begin{aligned}& \tilde{\mathcal{P}}\tilde{\mathcal{F}} + \tilde{\mathcal{F}}^T\tilde{\mathcal{P}} - \frac{\beta}{4}\lambda_2^{min}(\mathcal{L}^{\sigma(t)})\tilde{\mathcal{R}} \\ = & \begin{bmatrix} \gamma(\bar{P}(A+BG) + (A+BG)^T\bar{P}) & \gamma\bar{P}BG \\ \gamma G^T B^T \bar{P} & \Upsilon \end{bmatrix} \end{aligned} \tag{2.25}$$

where $\Upsilon = PA + A^T P - \frac{\beta}{4}\lambda_2^{min}(\mathcal{L}^{\sigma(t)})C^T C$ is marked. Based on (2.12) in Algorithm 2.1, one can conclude that Υ is a negative definite matrix. Then, it follows from Lemma 1.8 that

$$\tilde{\mathcal{P}}\tilde{\mathcal{F}} + \tilde{\mathcal{F}}^T\tilde{\mathcal{P}} - \frac{\beta}{4}\lambda_2^{min}(\mathcal{L}^{\sigma(t)})\tilde{\mathcal{R}} < 0. \tag{2.26}$$

It follows from (2.26) that (2.23) can be written as

$$\dot{V}_2 \leq W(\hat{\eta}) + \frac{\beta}{2}\sum_{i\in I[1,N]} ve^{-wt}. \tag{2.27}$$

Select the Lyapunov function as $V_3 = V_2 + \frac{\beta}{2w}\sum_{i\in I[1,N]} ve^{-wt}$. The derivative of V_3 can be calculated as $\dot{V}_3 \leq W(\hat{\eta}) \leq 0$. Taking advantage of LaSalle-Yoshizawa theorem in Krstić et al. (1995) that $\lim_{t\to\infty} W(\hat{\eta}) = 0$, which indicates that $\lim_{t\to\infty}\hat{\eta} = 0$. Furthermore, one obtains that $\lim_{t\to\infty}\eta = 0$. Therefore, we can obtain that the consensus error $\bar{\eta}$ asymptotically converges to zero and the adaptive weights d_{ij} converge to some finite values.

Finally, we prove that the Zeno behavior does not exist under the designed triggering mechanism. Since the switching instants satisfy that $t_{k+1} - t_k > \tau > 0$, so topology switches can't cause Zeno behavior. Thus, there are three cases to analyze.

Case 1: Consider the time interval $[t_{k'}^i, t_{k'+1}^i)$. When $t \in [t_{k'}^i, t_{k'+1}^i)$, calculate the derivative is obtained as

$$\begin{aligned}\dot{e}_i =& (I_2 \otimes A)e_i + \big((I_2 \otimes A) - \mathcal{F}\big)z_i \\ &- \sum_{j\in I[1,N]} d_{ij}a_{ij}^{\sigma(t)}\mathcal{M}(\hat{z}_i - \hat{z}_j).\end{aligned} \tag{2.28}$$

Therefore, the derivative of $\|e_i\|$ can be obtained that

$$\begin{aligned}\frac{d\|e_i\|}{dt} \leq& \|\dot{e}_i\| \\ \leq& \|I_2 \otimes A\|\|e_i\| + \|\big((I_2 \otimes A) - \mathcal{F}\big)z_i\| \\ &+ \|\sum_{j\in I[1,N]} d_{ij}a_{ij}^{\sigma(t)}\mathcal{M}(\hat{z}_i - \hat{z}_j)\|.\end{aligned} \tag{2.29}$$

According to the above analysis, we can obtain that there exists limited positive constant d such that $d_{ij} \leq d$ holds. According to Eq. (2.4) and matrix $A + BG$ being Hurwitz, one can obtain that ξ is bounded. Since the linear system does not have a finite escape time, according to the definition of z, z is bounded for any limited time. $e^{A(t-t_{k'}^i)}$ is bounded for any $t \in [t_{k'}^i, t_{k'+1}^i)$, so $\hat{z}(t)$ is bounded. Based on the above analysis, it is easy to obtain $\|\big((I_2 \otimes A) - \mathcal{F}\big)z_i\|$ and $\|\sum_{j\in I[1,N]} d_{ij}a_{ij}\mathcal{M}(\hat{z}_i - \hat{z}_j)\|$ are bounded. Moreover, there exists positive constant ϖ such that $\|\big((I_2 \otimes A) - \mathcal{F}\big)z_i\| + \|\sum_{j\in I[1,N]} d_{ij}a_{ij}^{\sigma(t)}\mathcal{M}(\hat{z}_i - \hat{z}_j)\| \leq \varpi$. Therefore, it follows from Eq. (2.29) that

$$\begin{aligned} \frac{d\|e_i\|}{dt} &\leq \|I_2 \otimes A\|\|e_i\| + \varpi \\ &= \|A\|\|e_i\| + \varpi. \end{aligned} \tag{2.30}$$

Define the function $y: [0, \infty) \to R_{\geq 0}$ such that

$$\begin{aligned} \dot{y}(t) &= \|A\|y(t) + \varpi, \\ y(0) &= \|e_i(t_{k'}^i)\| = 0. \end{aligned} \tag{2.31}$$

Using Comparison principle in Hassan (2012), one obtains

$$\|e_i\| \leq y(t - t_{k'}^i) \tag{2.32}$$

in which $y(t) = -\frac{\varpi}{\|A\|} + \frac{\varpi}{\|A\|}e^{\|A\|t}$ is the solution of (2.31). Therefore, one obtains that

$$\|e_i\| \leq \frac{\varpi}{\|A\|}e^{\|A\|(t-t_{k'}^i)} - \frac{\varpi}{\|A\|}. \tag{2.33}$$

Based on the triggering function (2.11), we can obtain that $h_i \leq 0$ if the following formula holds:

$$\|e_i\| \leq \sqrt{\frac{\nu e^{-wt}}{\sum\limits_{j\in I[1,N]} a_{ij}^{\sigma(t)}\|\mathcal{R}\|(1+\iota d)}}. \tag{2.34}$$

It is not difficult to observe that the lower bound of the time interval between two adjacent triggering instants $t_{k'}^i$ and $t_{k'+1}^i$ for subsystem i can be calculated by the time evolving from 0 to the right of (2.34). Hence, the lower bound ς_k^i of $t_{k'+1}^i - t_{k'}^i$ can be obtained that

$$-\frac{\varpi}{\|A\|} + \frac{\varpi}{\|A\|}e^{\|A\|\varsigma_k^i} \geq \sqrt{\frac{\nu e^{-w(t_{k'}^i+\varsigma_k^i)}}{\sum\limits_{j\in I[1,N]} a_{ij}(t)\|\mathcal{R}\|(1+\iota d)}}. \tag{2.35}$$

Furthermore, one obtains that

$$\varsigma_k^i \geq \frac{1}{\|A\|}ln\Bigg\{\sqrt{\frac{\nu e^{-w(t_{k'}^i+\varsigma_k^i)}}{\sum\limits_{j\in I[1,N]} a_{ij}(t)\|A\|(1+\iota d)}}\frac{\|A\|}{\varpi} + 1\Bigg\}. \tag{2.36}$$

It is obvious that $\varsigma_k^i > 0$ holds for any limited time.

Since the right-hand side of the inequality in Eq. (2.36) only approaches zero when time t approaches infinity. When t approaches infinity, one concludes $t^i_{k'}$ approaches infinity. Therefore, we can conclude that the multiagent system will not exhibit Zeno behavior when time satisfies $t \in [t^i_{k'}, t^i_{k'+1})$.

Case 2: Consider the time interval $[t_g, t^i_{k'+1})$ with $t^i_{k'+1} - t_g < \tau$. The topology graph is switching at time instant t_g.Hence, t_g may be seen as the new triggering instant $t^i_{k'}$, which means that Case 2 is equivalent to Case 1, so the proof process is omitted here.

Case 3: Consider the time interval $[t^i_{k'}, t_g)$ with $t_g - t^i_{k'} < \tau$. The next triggering interval is $[t_g, t^i_{k'+1})$. Since $t^i_{k'+1}$ comes not later than t_{g+1}. Thus, we only need to analyze that the minimum interaction interval $[t_m, t^i_{k'+1})$ is strictly positive. The exclusion of Zeno behavior is the same as the Case 2, so the proof process is omitted here. □

Remark 2.5 From Eq. (2.36), it can be seen that the lower bound of the triggering interval of adjacent events depends on the specific triggering time of events. Obviously, when the time tends to infinity, the exponential decay term tends to zero. Because the triggering mechanism ignores the state-dependent term and only analyzes the exponential decay term, the lower bound obtained is relatively conservative.

In light of Algorithm 2.1 and Theorem 2.1, we conclude that the stability of dynamics in Eq. (2.9) and the following dynamics is equal

$$\begin{cases} \dot{z}_i = \mathcal{F} z_i + \sum\limits_{j \in I[1,N]} d_{ij} a^{\sigma(t)}_{ij} \mathcal{M}(\hat{z}_i - \hat{z}_j) \\ \dot{d}_{ij} = k_{ij} a^{\sigma(t)}_{ij} (\hat{z}_i - \hat{z}_j)^T \mathcal{R}(\hat{z}_i - \hat{z}_j). \end{cases} \tag{2.37}$$

Remark 2.6 Note that according to the properties of the Laplace matrix, we can write dynamics in Eq. (2.37) as the following dynamics:

$$\begin{cases} \dot{z}_i = \mathcal{F} z_i + \sum\limits_{j \in I[1,N]} d_{ij} l^{\sigma(t)}_{ij} \mathcal{M} \hat{z}_j \\ \dot{d}_{ij} = k_{ij} a^{\sigma(t)}_{ij} (\hat{z}_i - \hat{z}_j)^T \mathcal{R}(\hat{z}_i - \hat{z}_j) \end{cases} \tag{2.38}$$

in which $l^{\sigma(t)}_{ij}$ expresses ith row and jth column of the matrix $\mathcal{L}^{\sigma(t)}$. Clearly dynamics in Eq. (2.38) can be seen as a linearized model of a complex network extended by the event-triggered mechanism in Li et al. (2010b). Theorem 2.1 shows that the complex network can realize network synchronization.

2.3.2 Leader-Follower Consensus

Note the leaderless consensus discussed in the previous subsection, whose consensus value cannot be directly determined by the system's dynamics, initial states, or topology structure. In order to enable the agents to track a predetermined signal, this subsection considers the leader-follower consensus problem.

Before presenting the main results, provide the following assumption.

Assumption 2.3 The topology $\bar{\mathcal{G}}^{\sigma(t)}$ is connected and undirected; moreover, there exists a directed spanning tree with the leader as the root node.

Under the condition of Assumption 2.3, the leader-follower matrix $\mathcal{H}^{\sigma(t)}$ of topology $\mathcal{G}^{\sigma(t)}$ is a positive definite matrix by Lemma 4 in Hu and Hong (2007).

Then, based on the system's output information, design the observer-based adaptive event-triggered consensus control strategy for follower i ($i \in I[1, N]$) as follows:

$$\begin{cases} u_i = G\xi_i \\ \dot{\varsigma}_i = A\varsigma_i + Bu_i + F(C\varsigma_i - y_i) \\ \dot{\xi}_i = (A + BG)\xi_i + EC \sum\limits_{j \in I[0,N]} d_{ij} a_{ij}^{\sigma(t)} \big((\hat{\xi}_i - \hat{\xi}_j) - (\hat{\varsigma}_i - \hat{\varsigma}_j)\big) \\ \dot{d}_{ij} = k_{ij} a_{ij}^{\sigma(t)} \begin{bmatrix} \hat{\varsigma}_i - \hat{\varsigma}_j \\ \hat{\xi}_i - \hat{\xi}_j \end{bmatrix}^T \mathcal{R} \begin{bmatrix} \hat{\varsigma}_i - \hat{\varsigma}_j \\ \hat{\xi}_i - \hat{\xi}_j \end{bmatrix} \end{cases} \tag{2.39}$$

in which ξ_i denotes the controller state, ς_i expresses the observer state, d_{ij} ($d_{ij}(0) = d_{ji}(0)$) denotes the adaptive coupling weight, $k_{ij} = k_{ji}$ are given positive constants, feedback gain matrices F, G, E, $\mathcal{R}$ are given in Algorithm 2.1, and $\hat{\xi}_i$, $\hat{\varsigma}_i$ are model-based estimators.

Note that $\xi_0 = 0$ and ς_0 in Eq. (2.39) are set as

$$\begin{cases} \dot{\varsigma}_0 = A\varsigma_0 + Bu_0 + F(C\varsigma_1 - y_0) \\ u_0 = G\xi_0 \end{cases} \tag{2.40}$$

in which ς_0 and ξ_0 respectively denote the observer state and controller state for the leader. What is meant to be noted is that $\hat{\xi}_0 = 0$ and $\hat{\varsigma}_0 = \varsigma_0$. Define observer error for the leader $\epsilon_0 = \varsigma_0 - x_0$.

For the convenience of writing, the following variables are provided, $v_i = \text{col}(\varsigma_i, \xi_i)$, $v = \text{col}(v_1, \ldots, v_N)$, $\bar{v}_i = \text{col}(x_i, \xi_i)$, $\bar{v} = \text{col}(\bar{v}_1, \ldots, \bar{v}_N)$, $\hat{v}_i = \text{col}(\hat{\varsigma}_i, \hat{\xi}_i)$, $\hat{v} = \text{col}(\hat{v}_1, \ldots, \hat{v}_N)$, $\zeta_i = v_i - v_0$, $\zeta = \text{col}(\zeta_1, \ldots, \zeta_N)$, $\bar{\zeta}_i = \bar{v}_i - v_1$, $\bar{\zeta} = \text{col}(\bar{\zeta}_1, \ldots, \bar{\zeta}_N)$.

Since $\lim_{t\to\infty} \epsilon_i = 0$ by Eq. (2.14), we have

$$\begin{cases} \dot{v}_0 = \mathcal{F} v_0 + \Xi \epsilon_0 \\ \dot{v}_i = \mathcal{F} v_i + \sum_{j\in I[0,N]} d_{ij} a_{ij}^{\sigma(t)} \mathcal{M}(\hat{v}_i - \hat{v}_j) + \Xi \epsilon_i \end{cases} \tag{2.41}$$

and

$$\begin{cases} \dot{\zeta}_i = \mathcal{F} \zeta_i + \sum_{j\in I[0,N]} d_{ij} a_{ij}^{\sigma(t)} \mathcal{M}(\hat{v}_i - \hat{v}_j) + \Xi(\epsilon_i - \epsilon_0) \\ \dot{d}_{ij} = k_{ij} a_{ij}^{\sigma(t)} (\hat{v}_i - \hat{v}_j)^T \mathcal{R}(\hat{v}_i - \hat{v}_j). \end{cases} \tag{2.42}$$

With the help of fact $\lim_{t\to\infty} \epsilon = 0$, decoupling from v, ζ yields that the stability of dynamics in Eqs. (2.39) and (2.40) is equivalent to the stability of dynamics in Eqs. (2.41) and (2.42). It follows from $\lim_{t\to\infty} \bar{\zeta} = 0$ that $\lim_{t\to\infty} x_i - x_0 = 0$. Thus, $\bar{\zeta}$ is regarded as the leader-follower consensus error. Obviously, it is easy to obtain $\lim_{t\to\infty} \zeta = 0 \Leftrightarrow \lim_{t\to\infty} \bar{\zeta} = 0$. Then, define the follower estimate error as $e_i = \hat{v}_i - v_i, \forall\, i = 1, \ldots, N$.

In this case, the triggering mechanism is the same as Eq. (2.6), the triggering function is constructed as

$$\begin{aligned} h_i = & \sum_{j\in I[1,N]} (2\iota d_{ij} + 1) a_{ij}^{\sigma(t)} e_i^T \mathcal{R} e_i + \iota d_{i0} a_{i0}^{\sigma(t)} e_i^T \mathcal{R} e_i \\ & - \frac{1}{4} \sum_{j\in I[1,N]} a_{ij}^{\sigma(t)} (\hat{v}_i - \hat{v}_j)^T \mathcal{R}(\hat{v}_i - \hat{v}_j) - \nu e^{-wt} \end{aligned} \tag{2.43}$$

in which $\iota,\ \nu,\ w$ are given positive constants.

Theorem 2.2 *Under the condition of Assumptions 2.1–2.3, the observer-based adaptive event-triggered control strategy* (2.39)–(2.40) *and triggering function* (2.43) *are constructed by Algorithm 2.1, feedback matrices F, G, E, and $\mathcal{R}$ are set such that $A + FC$ and $A + BG$ are Hurwitz matrices, $\mathcal{R} = \begin{bmatrix} C^TC & -C^TC \\ -C^TC & C^TC \end{bmatrix}$ and $E = -P^{-1}C^T$. Then, the leader-follower consensus can be reached, and adaptive coupling weights d_{ij} converge to finite values. Moreover, the MAS does not exhibit the Zeno behavior.*

Proof Consider the Lyapunov function candidate as

$$\begin{aligned} V_4 = & \frac{1}{2} \sum_{i\in I[1,N]} \zeta_i^T \mathcal{P} \zeta_i + \sum_{i\in I[1,N]} \sum_{j\in I[1,N]} \frac{(d_{ij} - \beta)^2}{8k_{ij}} \\ & + \sum_{i\in I[1,N]} \frac{(d_{i1} - \beta)^2}{4k_{i1}} \end{aligned} \tag{2.44}$$

in which $\beta = \max\{\frac{4}{\lambda_1^{min}(\mathcal{H}^{\sigma(t)})}, \frac{1}{\iota}\}$, $\lambda_1(\mathcal{H}^{\sigma(t)})$ denotes the smallest eigenvalue of $\mathcal{H}^{\sigma(t)}$. Calculate the derivative of Eq. (2.44) as

$$\begin{aligned}
\dot{V}_4 = & \sum_{i\in I[1,N]} \zeta_i^T \mathcal{P}\dot{\zeta}_i + \sum_{i\in I[1,N]} \sum_{j\in I[1,N]} \frac{(d_{ij}-\beta)}{4k_{ij}}\dot{d}_{ij} \\
& + \sum_{i\in I[1,N]} \frac{(d_{i0}-\beta)}{2k_{i0}}\dot{d}_{i0} \\
= & \sum_{i\in I[1,N]} \zeta_i^T \mathcal{P}\mathcal{F}\zeta_i - \sum_{i\in I[1,N]} \sum_{j\in I[1,N]} d_{ij}a_{ij}^{\sigma(t)}\zeta_i^T \mathcal{R}(\hat{v}_i - \hat{v}_j) \\
& - \sum_{i\in I[1,N]} d_{i0}a_{i0}^{\sigma(t)}\zeta_i^T \mathcal{R}(\hat{v}_i - \hat{v}_0) \\
& + \sum_{i\in I[1,N]} \sum_{j\in I[1,N]} \frac{d_{ij}-\beta}{4} a_{ij}^{\sigma(t)} (\hat{v}_i - \hat{v}_j)^T \mathcal{R}(\hat{v}_i - \hat{v}_j) \\
& + \sum_{i\in I[1,N]} \frac{d_{i0}-\beta}{2} a_{i0}^{\sigma(t)} (\hat{v}_i - \hat{v}_0)^T \mathcal{R}(\hat{v}_i - \hat{v}_0).
\end{aligned} \tag{2.45}$$

Taking advantage of Lemma 1.6 yields

$$\begin{aligned}
& - \sum_{i\in I[1,N]} \sum_{j\in I[1,N]} d_{ij}a_{ij}^{\sigma(t)}\zeta_i^T \mathcal{R}(\hat{v}_i - \hat{v}_j) \\
\le & -\frac{1}{4} \sum_{i\in I[1,N]} \sum_{j\in I[1,N]} d_{ij}a_{ij}^{\sigma(t)} (\hat{v}_i - \hat{v}_j)^T \mathcal{R}(\hat{v}_i - \hat{v}_j) \\
& + \sum_{i\in I[1,N]} \sum_{j\in I[1,N]} d_{ij}a_{ij}^{\sigma(t)} e_i^T \mathcal{R} e_i.
\end{aligned} \tag{2.46}$$

It follows from $\zeta_i = (\hat{v}_i - \hat{v}_1) - e_i$ that

$$\begin{aligned}
& - \sum_{i\in I[1,N]} d_{i0}a_{i0}^{\sigma(t)}\zeta_i^T \mathcal{R}(\hat{v}_i - \hat{v}_0) \\
= & - \sum_{i\in I[1,N]} d_{i0}a_{i0}^{\sigma(t)} (\hat{v}_i - \hat{v}_0)^T \mathcal{R}(\hat{v}_i - \hat{v}_0) \\
& + \sum_{i\in I[1,N]} d_{i0}a_{i0}^{\sigma(t)} e_i^T \mathcal{R}(\hat{v}_i - \hat{v}_0) \\
\le & -\frac{1}{2} \sum_{i\in I[1,N]} d_{i0}a_{i0}^{\sigma(t)} (\hat{v}_i - \hat{v}_0)^T \mathcal{R}(\hat{v}_i - \hat{v}_0) \\
& + \frac{1}{2} \sum_{i\in I[1,N]} d_{i0}a_{i0}^{\sigma(t)} e_i^T \mathcal{R} e_i.
\end{aligned} \tag{2.47}$$

Taking Eqs. (2.46) and (2.47) into Eq. (2.45) that

$$\dot{V}_4 \leq \sum_{i\in I[1,N]} \zeta_i^T \mathcal{P}\mathcal{F}\zeta_i - \frac{\beta}{4}\sum_{i\in I[1,N]}\sum_{j\in I[1,N]} a_{ij}^{\sigma(t)}(\hat{v}_i - \hat{v}_j)^T \mathcal{R}(\hat{v}_i - \hat{v}_j) + \sum_{i\in I[1,N]}\sum_{j\in I[1,N]} d_{ij}a_{ij}^{\sigma(t)} e_i^T \mathcal{R} e_i + \frac{1}{2}\sum_{i\in I[1,N]} d_{i0}a_{i0}^{\sigma(t)} e_i^T \mathcal{R} e_i. \tag{2.48}$$

Based on the fact that $\hat{v}_i - \hat{v}_j = e_i - e_j + \zeta_i - \zeta_j$ and Lemma 1.6, we can obtain that

$$\begin{aligned} &-\frac{\beta}{8}\sum_{i\in I[1,N]}\sum_{j\in I[1,N]} a_{ij}^{\sigma(t)}(\hat{v}_i - \hat{v}_j)^T \mathcal{R}(\hat{v}_i - \hat{v}_j) \\ &\leq -\frac{\beta}{8}\sum_{i\in I[1,N]}\sum_{j\in I[1,N]} a_{ij}^{\sigma(t)} \zeta_i^T \mathcal{R}(\zeta_i - \zeta_j) \\ &\quad + \frac{\beta}{2}\sum_{i\in I[1,N]}\sum_{j\in I[1,N]} a_{ij}^{\sigma(t)} e_i^T \mathcal{R} e_i. \end{aligned} \tag{2.49}$$

Then, based on Eq. (2.49), we can obtain that Eq. (2.48) can be written as

$$\begin{aligned} \dot{V}_4 \leq &\sum_{i\in I[1,N]} \zeta_i^T \mathcal{P}\mathcal{F}\zeta_i - \frac{\beta}{8}\sum_{i\in I[1,N]}\sum_{j\in I[1,N]} a_{ij}^{\sigma(t)} \zeta_i^T \mathcal{R}(\zeta_i - \zeta_j) \\ &+ \frac{\beta}{2}\sum_{i\in I[1,N]}\Big(\frac{\iota}{\beta\iota} d_{i0}a_{i0}^{\sigma(t)} e_i^T \mathcal{R} e_i + \sum_{j\in I[1,N]} (1 + \frac{2\iota}{\beta\iota} d_{ij}) a_{ij}^{\sigma(t)} e_i^T \mathcal{R} e_i \\ &- \frac{1}{4}\sum_{j\in I[1,N]} a_{ij}^{\sigma(t)}(\hat{v}_i - \hat{v}_j)\mathcal{R}(\hat{v}_i - \hat{v}_j)\Big). \end{aligned} \tag{2.50}$$

Based on triggering mechanism (2.6) and triggering function (2.43), $\beta = \max\{\frac{4}{\lambda_1^{min}(\mathcal{L}_2^{\sigma(t)})}, \frac{1}{\iota}\}$, we can obtain that

$$\begin{aligned} \dot{V}_4 \leq &\sum_{i\in I[1,N]} \zeta_i^T \mathcal{P}\mathcal{F}\zeta_i - \frac{\beta}{8}\sum_{i\in I[1,N]}\sum_{j\in I[1,N]} a_{ij}^{\sigma(t)} \zeta_i^T \mathcal{R}(\zeta_i - \zeta_j) \\ &+ \frac{\beta}{2}(N-1)\nu e^{-wt}. \end{aligned} \tag{2.51}$$

Introducing the following variables and transformations $U = \begin{bmatrix} I & 0 \\ -I & I \end{bmatrix}$, $\tilde{\zeta}_i = U\zeta_i$, $\tilde{\mathcal{P}} = U^{-T}\mathcal{P}U^{-1}$, $\tilde{\mathcal{F}} = U\mathcal{F}U^{-1}$, $\tilde{\mathcal{R}} = U^{-T}\mathcal{R}U^{-1}$. Then, we can obtain

$$\sum_{i\in I[1,N]} \zeta_i^T \mathcal{PF}\zeta_i - \frac{\beta}{8}\sum_{i\in I[1,N]}\sum_{j\in I[1,N]} a_{ij}(t)\zeta_i^T \mathcal{R}(\zeta_i - \zeta_j)$$
$$\leq \frac{1}{2}\tilde{\zeta}^T\left(I_N \otimes (\tilde{\mathcal{P}}\tilde{\mathcal{F}} + \tilde{\mathcal{F}}^T\tilde{\mathcal{P}}) - \frac{\beta}{4}\mathcal{H}^{\sigma(t)} \otimes \tilde{\mathcal{R}}\right)\tilde{\zeta}. \tag{2.52}$$

Since $\mathcal{H}^{\sigma(t)}$ is the positive definite matrix, there must exist a unitary matrix $J^{\sigma(t)}$ satisfying $J^{\sigma(t)}\mathcal{H}^{\sigma(t)}(J^{\sigma(t)})^T = \left\{\lambda_1(\mathcal{H}_2^{\sigma(t)}), \ldots, \lambda_N(\mathcal{H}_2^{\sigma(t)})\right\} \triangleq \Omega^{\sigma(t)}$, where $0 < \lambda_1(\mathcal{H}_2^{\sigma(t)}) \leq \cdots \leq \lambda_N(\mathcal{H}_2^{\sigma(t)})$ holds.

Introducing the variable $\hat{\zeta} = (J^{\sigma(t)} \otimes I_{2n})\tilde{\zeta}$, we can obtain that Eq. (2.52) satisfies

$$\sum_{i\in I[2,N]} \zeta_i^T \mathcal{PF}\zeta_i - \frac{\beta}{8}\sum_{i\in I[2,N]}\sum_{j\in I[2,N]} a_{ij}^{\sigma(t)}\zeta_i^T \mathcal{R}(\zeta_i - \zeta_j)$$
$$= \frac{1}{2}\hat{\zeta}^T\left\{I_{N-1} \otimes (\tilde{\mathcal{P}}\tilde{\mathcal{F}} + \tilde{\mathcal{F}}^T\tilde{\mathcal{P}}) - \frac{\beta}{4}(\Omega^{\sigma(t)} \otimes \tilde{\mathcal{R}})\right\}\hat{\zeta} \tag{2.53}$$
$$\leq \frac{1}{2}\sum_{i\in I[2,N]} \hat{\zeta}_i^T\left(\tilde{\mathcal{P}}\tilde{\mathcal{F}} + \tilde{\mathcal{F}}^T\tilde{\mathcal{P}} - \frac{\beta}{4}\lambda_1^{min}(\mathcal{H}^{\sigma(t)})\tilde{\mathcal{R}}\right)\hat{\zeta}_i.$$

Similar to Theorem 2.1, we conclude $\lim_{t\to\infty}\hat{\zeta} = 0$, which further shows $\lim_{t\to\infty}\zeta = 0$ and the adaptive coupling weights d_{ij} converge to some finite values. In addition, the Zeno behavior is excluded from the proposed event-triggered control protocol, and its proof process is omitted. □

2.4 Consensus Under Communication Link Faults

The previous section only considered the consensus problem of multi-agent systems under undirected switching topology; however, the actual system may suffer from communication failures. Communication failures can affect the safety and stability of the system, leading to a decrease in system control performance. In addition, undirected topology is a special case of directed topology. Therefore, this section focuses on the leader-follower consensus under communication link faults in directed topology.

2.4.1 Communication Link Fault Model

The communication link faults for digraph considered in this section are modeled as

$$\bar{a}_{ij}(t) = a_{ij} + \psi_{ij}^f(t)$$
$$\bar{a}_{i0}(t) = a_{i0} + \psi_{i0}^f(t),\ i, j \in I[1, N], \tag{2.54}$$

where a_{ij} and a_{i0} denote the weights for the edge (i, j) and the edge $(i, 0)$ in the ideal communication topology. $\psi_{ij}^{f}(t)$ and $\psi_{i0}^{f}(t)$ denote the corrupted weights caused by communication link faults for the edge (i, j) and the edge $(i, 0)$.

In this case, the corresponding Laplace matrix of digraph $\bar{\mathcal{G}}$ is denoted by $\mathcal{L}(t) = \Delta(t) - \mathcal{A}(t)$, where $\Delta(t) = \text{diag}\{\sum_{j\in I[1,N]} a_{1j}(t) \cdots \sum_{j\in I[1,N]} a_{Nj}(t)\}$ and $\mathcal{A}(t) = [\bar{a}_{ij}(t)]$ and $\mathcal{A}_0(t) = [\bar{a}_{i0}(t)]$. The corresponding leader-follower matrix $\mathcal{H}(t)$ is denoted as $\mathcal{H}(t) = \mathcal{L}(t) + \mathcal{A}_0(t)$.

To solve the leader-follower consensus problem for digraph $\bar{\mathcal{G}}$ under communication link faults, the following essential assumptions are given.

Assumption 2.4 The digraph $\bar{\mathcal{G}}$ with communication link faults exists a spanning tree with the leader being root node.

Assumption 2.5 The corrupted weights $\psi_{ij}^{f}(t)$, $\psi_{i0}^{f}(t)$ for communication link faults and their derivatives $\dot{\psi}_{ij}^{f}(t)$, $\dot{\psi}_{i0}^{f}(t)$ are bounded.

Assumption 2.6 The symbols of a_{ij} and a_{i0} are the same as those of $\bar{a}_{ij}(t)$ and $\bar{a}_{i0}(t)$, respectively.

Remark 2.7 Assumptions 2.4–2.6 are standard and mild, which can be found in Li et al. (2015) and Chen et al. (2020). Assumption 2.4 guarantees that when there exist communication link faults, the connectivity of the graph is the same as if the graph does not fail.

Lemma 2.2 Chen et al. 2020 *Under condition of Assumptions 2.4–2.6, the real part of all the eigenvalues of* $\mathcal{H}(t)$ *is greater than zero, there exists a positive diagonal matrix* $Q(t) = diag\{q_1, \ldots, q_N\}$ *with* $Q(t)\mathcal{H}(t) + \mathcal{H}(t)^T Q(t)$ *being a positive definite matrix. Especially,* $q(t) = col(q_1(t), \ldots, q_N(t)) = (\mathcal{H}(t))^{-1}1_N$. *In addition,* $Q(t)$ *and* $\dot{Q}(t)$ *are norm bounded.*

2.4.2 Controller Design Under Link Faults

To quantify the control objective, construct the local consensus error for agent i as $\tilde{\epsilon}_i = x_i - x_0$, $i \in I[1, N]$. Then, define the global consensus error $\tilde{\epsilon} = \text{col}(\tilde{\epsilon}_1, \ldots, \tilde{\epsilon}_N)$. Thus, the goal of leader-follower consensus under communication link faults is achieved if $\lim_{t\to\infty} \tilde{\epsilon} = 0$.

In order to overcome the influence of communication link faults on system stability, the adaptive event-triggered observer is designed, and the leader-follower consensus protocol is combined to realize the leader-follower consensus of multi-agent systems under communication link faults.

Firstly, design the following adaptive event-triggered observer:

$$
\begin{aligned}
\dot{\zeta}_i &= A\zeta_i - \left(c_i(t_k^i) + \varrho_i(t_k^i)\right) P\xi_i(t_k^i) \\
\varrho_i &= \xi_i^T P\xi_i
\end{aligned}
$$

$$\dot{c}_i = \xi_i(t_k^i)^T P^2 \xi_i(t_k^i), \tag{2.55}$$

where ζ_i denotes the observer state, c_i denotes the adaptive gain with $c_i(0) \geq \max\{1, \sqrt{2\iota_1}\}$, $\iota_1 > 0$ and $\xi_i = \sum_{j \in I[1,N]} \bar{a}_{ij}(\zeta_i - \zeta_j) + \bar{a}_{i0}(\zeta_i - x_0)$ denotes the distributed leader state error, matrix P is to be given.

Define the global leader state estimation error for follower i as $\epsilon_i = \zeta_i - x_0$, $i \in I[1, N]$. For the convenience of writing, we give the compact set form of the following variables. $\zeta = \text{col}(\zeta_1, \ldots, \zeta_N)$, $\tilde{c} = \text{diag}\{c_1(t_k^1), \ldots, c_N(t_k^1)\}$, $\tilde{\varrho} = \text{diag}\{\varrho_1(t_k^1), \ldots, \varrho_N(t_k^N)\}$, $c = \text{diag}\{c_1, \ldots, c_N\}$, $\varrho = \text{diag}\{\varrho_1, \ldots, \varrho_N\}$, $\xi = \text{col}(\xi_1, \ldots, \xi_N)$, $\epsilon = \text{col}(\epsilon_1, \ldots, \varrho_N)$, $x = \text{col}(x_1, \ldots, x_N)$.

Thus, we can obtain that $\xi = \big(\mathcal{H}(t) \otimes I\big)(\zeta - 1_N \otimes x_0)$ and $\dot{\zeta} = (I \otimes A)\zeta - \big((\tilde{c} + \tilde{\varrho}) \otimes P\big)\tilde{\xi}$, $\|\epsilon\| \leq \|\mathcal{H}^{-1}(t)\| \|\xi\|$. Furthermore, we calculate the derivative of ξ as follows,

$$\dot{\xi} = (\dot{\mathcal{H}}(t) \otimes I)\epsilon + (I \otimes A)\xi - \big(\mathcal{H}(t)(\tilde{c} + \tilde{\varrho}) \otimes P\big)\tilde{\xi}. \tag{2.56}$$

Define the error variables before constructing triggering mechanism as follows:

$$e_i = \xi_i(t_k^i) - \xi_i,\ e_{\varrho_i} = \varrho_i(t_k^i) - \varrho_i,\ e_{c_i} = c_i(t_k^i) - c_i,\ t \in [t_k^i, t_{k+1}^i).$$

Then, construct the following triggering mechanism:

$$t_{k+1}^i = \inf\{t > t_k^i | g_{1i} \geq 0 \vee g_{2i} \geq 0\}, \tag{2.57}$$

in which t_k^i is the triggering instant with $t_0^i = 0$, g_{1i} and g_{2i} are triggering functions constructed by

$$\begin{aligned} g_{1i} =& -\iota_1 \xi_i^T P^2 \xi_i - \vartheta_{1i} e^{-\pi_{1i} t} + (c_i + \varrho_i)^2 e_i^T P^2 e_i \\ g_{2i} =& -\iota_2 \xi_i^T P^2 \xi_i - \vartheta_{2i} e^{-\pi_{2i} t} + (e_{c_i} + e_{\varrho_i})^2 \tilde{\xi}_i^T P^2 \tilde{\xi}_i \end{aligned} \tag{2.58}$$

with ι_j, ϑ_{ji}, π_{ji} being positive constants and $0 < \iota_1 < 2$ for $j = 1, 2$, $i \in I[1, N]$.

Remark 2.8 This section proposes a hybrid event-triggered mechanism to address the limited communication resources issue. When the triggering mechanism is met, i.e., $g_{1i} \geq 0$ or $g_{2i} \geq 0$, the measurement errors are reset as $e_i(t_k^i) = 0$, $e_{c_i}(t_k^i) = 0$, $e_{\varrho_i}(t_k^i) = 0$. At this time, $g_{1i} < 0$ or $g_{2i} < 0$.

Next, we will present the feasibility of designing the observer in Eq. (2.55).

Theorem 2.3 *Under the condition of Assumptions 2.4–2.6, the adaptive event-triggered observer in Eq.* (2.55) *can converge to leader state x_0 under the designed triggering mechanism in Eq.* (2.57) *with P being the solution of the following equation:*

$$A^T P + PA - P^2 + I = 0. \tag{2.59}$$

In addition, the Zeno behavior is excluded under the designed triggering mechanism in Eq. (2.57).

Proof Consider the following Lyapunov function candidate:

$$V_5 = \frac{1}{2} \sum_{i \in I[1,N]} \nu_i (2c_i + \varrho_i) q_i(t) \varrho_i + \sum_{i \in I[1,N]} (c_i - c_\beta)^2, \tag{2.60}$$

where the parameter c_β is a positive constant to be given and $q_i(t)$ is given in Lemma 2.2. Then, calculate the derivative of V_5

$$\begin{aligned}
\dot{V}_5 =& \frac{1}{2} \sum_{i \in I[1,N]} (2\dot{c}_i + \dot{\varrho}_i) q_i(t) \varrho_i + \frac{1}{2} \sum_{i \in I[1,N]} (2c_i + \varrho_i)\big(\dot{q}_i(t)\varrho_i + q_i(t)\dot{\varrho}_i\big) \\
&+ 2 \sum_{i \in I[1,N]} (c_i - c_\beta) \tilde{\xi}_i^T P^2 \tilde{\xi}_i \\
\leq& 2 \sum_{i \in I[1,N]} (c_i + \varrho_i) q_i(t) \xi_i^T P \dot{\xi}_i + \sum_{i \in I[1,N]} q_i(t) \varrho_i \tilde{\xi}_i^T P^2 \tilde{\xi}_i \\
&+ \frac{1}{2} \sum_{i \in I[1,N]} (2c_i + \varrho_i) \dot{q}_i(t) \xi_i^T P \xi_i + \frac{2}{q_m} \sum_{i \in I[1,N]} q_i(t)(c_i - c_\beta) \tilde{\xi}_i^T P^2 \tilde{\xi}_i \\
=& 2\xi^T \big((c + \varrho) Q(t) \otimes P\big) \dot{\xi} + \tilde{\xi}^T (\varrho Q(t) \otimes P^2) \tilde{\xi} + \frac{1}{2} \xi^T \big((2c + \varrho) \dot{Q}(t) \otimes P\big) \xi \\
&+ \frac{2}{q_m} \tilde{\xi}^T \big((c - c_\beta I) Q(t) \otimes P^2\big) \tilde{\xi} \\
=& 2\xi^T \big((c + \varrho) Q(t) \dot{\mathcal{H}}(t) \otimes Q\big) \epsilon + \xi^T \big((c + \varrho) Q(t) \otimes (PA + A^T P)\big) \xi \\
&- \xi^T \big((c + \varrho)(Q(t)\mathcal{H}(t) + \mathcal{H}(t)^T Q(t))(\tilde{c} + \tilde{\varrho}) \otimes P^2\big) \tilde{\xi} \\
&+ \tilde{\xi}^T (\varrho Q(t) \otimes P^2) \tilde{\xi} + \frac{1}{2} \xi^T \big((2c + \varrho) \dot{Q}(t) \otimes P\big) \xi \\
&+ \frac{2}{q_m} \tilde{\xi}^T (c - c_\beta I) Q(t) \otimes P^2) \tilde{\xi},
\end{aligned} \tag{2.61}$$

where $q_m = \min_{\{\forall\, t \geq 0,\ i \in\, I[1,N]\}} \{q_i(t)\}$ is defined. Considering the following formula:

$$\begin{aligned}
& - \xi^T \big((c + \varrho)(Q(t)\mathcal{H}(t) + \mathcal{H}(t)^T Q(t))(\tilde{c} + \tilde{\varrho}) \otimes P^2\big) \tilde{\xi} \\
=& - \xi^T \big((c + \varrho) \bar{\mathcal{H}}(t)(c + \varrho) \otimes P^2\big) \xi \\
& - 2\xi^T \big((c + \varrho) Q(t) \mathcal{H}(t)(c + \varrho) \otimes P^2\big) e \\
& - 2\xi^T \big((c + \varrho) Q(t) \mathcal{H}(t)(e_c + e_\varrho) \otimes P^2\big) \tilde{\xi} \\
\leq& - \lambda_m(\bar{\mathcal{H}}(t)) \xi^T \big((c + \varrho)^2 \otimes P^2\big) \xi \\
& + (o_1 + o_2) \lambda_M \tilde{\mathcal{H}}(t) \xi^T \big((c + \varrho)^2 \otimes P^2\big) \xi \\
& + \frac{1}{o_1} e^T \big((c + \varrho)^2 \otimes P^2\big) e + \frac{1}{o_2} \tilde{\xi}^T \big((e_c + e_\varrho)^2 \otimes P^2\big) \tilde{\xi}
\end{aligned}$$

$$
\begin{aligned}
&= - o_3\xi^T\big((c+\varrho)^2 \otimes P^2\big)\xi + \frac{1}{o_1}e^T\big((c+\varrho)^2 \otimes P^2\big)e \\
&\quad + \frac{1}{o_2}\tilde{\xi}^T\big((e_c+e_\varrho)^2 \otimes P^2\big)\tilde{\xi},
\end{aligned} \tag{2.62}
$$

where $o_3 = \lambda_m\big(\bar{\mathcal{H}}(t)\big) - (o_1+o_2)\lambda_M\big(\tilde{\mathcal{H}}(t)\big)$ is defined, $\lambda_M\big(\tilde{\mathcal{H}}(t)\big) = \lambda_{max}\big(Q(t)\mathcal{H}(t)\mathcal{H}(t)^TQ(t)\big)$, $\lambda_m\big(\bar{\mathcal{H}}(t)\big) = \lambda_{min}\{Q(t)\mathcal{H}(t) + \mathcal{H}(t)^TQ(t)\}$, and $o_j > 0$ for $j = 1, 2, 3$.

Furthermore, it follows from Lemma 1.6 that

$$
\begin{aligned}
&\tilde{\xi}^T\big(\varrho Q(t) \otimes P^2\big)\tilde{\xi} + \frac{2}{q_{min}}\tilde{\xi}^T\big((c - c_\beta I)Q(t) \otimes P^2\big)\tilde{\xi} \\
\le\,& \alpha_1\tilde{\xi}^T\big((c+\varrho)Q(t) \otimes P^2\big)\tilde{\xi} - \frac{2}{q_m}\tilde{\xi}^T\big(c_\beta Q(t) \otimes P^2\big)\tilde{\xi} \\
\le\,& 2\alpha_1\xi^T\big((c+\varrho)Q(t) \otimes P^2\big)\xi + 2\alpha_1 e^T\big((c+\varrho)Q(t) \otimes P^2\big)e \\
&+ \frac{2}{q_m}\Big(e^T\big(c_\beta Q(t) \otimes P^2\big)e - \frac{1}{2}\xi^T\big(c_\beta Q(t) \otimes P^2\big)\xi\Big),
\end{aligned} \tag{2.63}
$$

where $\alpha_1 = \max\{1, \frac{2}{q_m}\}$ is defined.

It follows from Lemma 1.6 that

$$
\begin{aligned}
&2\xi^T\big((c+\varrho)Q(t)\dot{\mathcal{H}}(t) \otimes P\big)\epsilon \\
\le\,& \delta_1 q_M^2\xi^T\big((c+\varrho)^2 \otimes P^2\big)\xi + \frac{\lambda_M\big(\dot{\mathcal{H}}(t)^T\dot{\mathcal{H}}(t)\big)}{\delta_1}\epsilon^T\epsilon,
\end{aligned} \tag{2.64}
$$

and

$$
\frac{1}{2}\xi^T\big((2c+\varrho)\dot{Q}(t) \otimes P\big)\xi \le \frac{\dot{q}_M}{\lambda_m(P)}\xi^T\big((c+\varrho) \otimes P^2\big)\xi, \tag{2.65}
$$

where $\delta_1 > 0$, $q_M^2 = \max_{\{\forall\, t\ge 0, i\in\, I[1,N]\}}\{q_i^2(t)\}$, $\dot{q}_M = \max_{\{\forall\, t\ge 0, i\in\, I[1,N]\}}\{\dot{q}_i(t)\}$ are defined.

Substituting the above inequalities into Eq. (2.61) that

$$
\begin{aligned}
\dot{V}_5 \le\,& \xi^T\big((c+\varrho)Q(t) \otimes (PA + A^TP)\big)\xi - o_3\xi^T\big((c+\varrho)^2 \otimes P^2\big)\xi \\
&+ \frac{1}{o_1}e^T\big((c+\varrho)^2 \otimes P^2\big)e + \frac{1}{o_2}\tilde{\xi}^T\big((e_c+e_\varrho)^2 \otimes P^2\big)\tilde{\xi} \\
&+ 2\alpha_1\xi^T\big((c+\varrho)Q(t) \otimes P^2\big)\xi + 2\alpha_1 e^T\big((c+\varrho)Q(t) \otimes P^2\big)e \\
&+ \frac{2}{q_m}\Big(e^T\big(c_\beta Q(t) \otimes P^2\big)e - \frac{1}{2}\xi^T\big(c_\beta Q(t) \otimes P^2\big)\xi\Big)
\end{aligned}
$$

$$+\delta_1 q_M^2 \xi^T\big((c+\varrho)^2 \otimes P^2\big)\xi + \frac{\lambda_M\big(\dot{\mathcal{H}}(t)\dot{\mathcal{H}}(t)\big)}{\delta_1}\epsilon^T\epsilon$$
$$+\frac{\dot{q}_M}{\lambda_m(P)}\xi^T\big((c+\varrho)\otimes P^2\big)\xi. \tag{2.66}$$

In addition, the following inequalities can be obtained through the triggering mechanism in Eq. (2.57):

$$e^T\big((c+\varrho)^2 \otimes P^2\big)e \le \iota_1\xi^T(I\otimes P^2)\xi + \sum_{i=1}^N \vartheta_{1i}e^{-\pi_{1i}t}$$
$$\tilde{\xi}^T\big((e_c+e_\varrho)^2 \otimes P^2\big)\tilde{\xi} \le \iota_2\xi^T(I\otimes P^2)\xi + \sum_{i=1}^N \vartheta_{2i}e^{-\pi_{2i}t}$$
$$e^T\big((c+\varrho)Q(t) \otimes P^2\big)e \le \frac{q_m}{c_m}\big(\iota_1\xi^T(I\otimes P^2)\xi + \sum_{i=1}^N \vartheta_{1i}e^{-\pi_{1i}t}\big)$$
$$e^T\big(c_\beta Q(t) \otimes P^2\big)e \le \frac{q_m c_\beta}{c_m^2}\big(\iota_1\xi^T(I\otimes P^2)\xi + \sum_{i=1}^N \vartheta_{1i}e^{-\pi_{1i}t}\big).$$

Substituting the above inequalities into Eq. (2.66) that

$$\dot{V}_5 \le \lambda_M\big(Q(t)\big)\xi^T\big((c+\varrho)\otimes(PA+A^TP)\big)\xi + \xi^T\Big(\big(-\omega_1(c+\varrho)^2$$
$$+\omega_2 I + \omega_3(c+\varrho) - \omega_4 I\big)\otimes P^2\Big) + \phi(t), \tag{2.67}$$

where $\omega_1 = o_3 - \delta_1 q_M^2 > 0$ can be ensured by selecting parameter $\delta_1 > 0$, $\omega_2 = \frac{\iota_1}{o_1} + \frac{\iota_2}{o_2} + \frac{2\alpha_1 q_m \iota_1}{c_m}$, $\omega_3 = 2\alpha_1\lambda_M\big(Q(t)\big) + \frac{\dot{q}_M}{\lambda_m(P)}$, $\omega_4 = \Big((1-\frac{2}{\iota_1})c_\beta - \frac{\lambda_M\big(\dot{\mathcal{H}}^T(t)\dot{\mathcal{H}}(t)\big)\|\mathcal{H}^{-1}(t)\|^2}{\delta_1\lambda_m(P^2)}\Big) > 0$ can be ensured by selecting sufficiently large c_β, $\phi(t) = (\frac{1}{o_1} + \frac{2\alpha_1 q_m}{c_m} + \frac{2c_\beta}{c_m^2})\sum_{i\in I[1,N]}\vartheta_{1i}e^{-\pi_{1i}t} + \frac{1}{o_2}\sum_{i\in I[1,N]}\vartheta_{2i}e^{-\pi_{2i}t}$. Note that $\|\epsilon\| \le \|\mathcal{H}^{-1}(t)\|\|\xi\|$ is used to obtain Eq. (2.67).

Setting $\omega_5 = 1 - \frac{2}{\iota_2}$, one can obtain the following inequality:

$$-\omega_1(c+\varrho)^2 + \omega_2 I + \omega_3(c+\varrho) - \omega_4 I \le -\lambda_M\big(Q(t)\big)(c+\varrho),$$

if $c_\beta \ge \frac{1}{\omega_5}\Big(\frac{\lambda_M\big(\dot{\mathcal{H}}(t)\dot{\mathcal{H}}(t)\big)\|\mathcal{H}^{-1}(t)\|^2}{\delta_1\lambda_m(P^2)} + \frac{\big(\omega_4+\lambda_M\big(Q(t)\big)\big)}{4\omega_1}\Big)$ is selected.

Substituting the above inequality into Eq. (2.67) yields

$$\dot{V}_5 \le \lambda_M\big(Q(t)\big)\xi^T\big((c+\varrho)\otimes(PA+A^TP-P^2)\big)\xi + \phi(t)$$
$$= -\lambda_M\big(Q(t)\big)\xi^T\big((c+\varrho)\otimes I\big)\xi + \phi(t). \tag{2.68}$$

Based on the definition of $\phi(t)$, one can conclude that $\lim_{t\to\infty}\phi(t)=0$. Set $W=V+(\frac{1}{o_1}+\frac{2\alpha_1 q_m}{c_m}+\frac{2c_\beta}{c_m^2})\sum_{i\in I[1,N]}\frac{\vartheta_{1i}}{\pi_{1i}}e^{-\pi_{1i}t}+\sum_{i\in I[1,N]}\frac{\vartheta_{2i}}{o_2\pi_{2i}}e^{-\pi_{1i}t}$. Calculate the derivative of W

$$\dot{W}\le-\lambda_M\big(Q(t)\big)\xi^T\big((c+\varrho)\otimes I\big)\xi.$$

Taking advantage of Lemma 1.5, we can obtain that $\lim_{t\to\infty}\xi=0$, which indicates that the adaptive event-triggered observer in Eq. (2.55) can estimate the leader state for each follower under the communication link faults.

Finally, we analyze that the system will not exhibit Zeno behavior. Let's assume that follower i exhibites the Zeno behavior, i.e., $\lim_{t\to\infty}t_k^i=T<\infty$. Based on the working principle of the triggering mechanism, we have $e_i(t_k^i)=0,\ e_{c_i}(t_k^i)=0,\ e_{\varrho_i}(t_k^i)=0$, either of the following formulas satisfy

$$\begin{aligned}&\big(c_i(t_k^i)+\varrho_i(t_k^i)\big)^2e_i^T(t_k^i)P^2e_i(t_k^i)\ge\vartheta_{1i}e^{-\pi_{1i}t_k^i}\\&\big(e_{c_i}(t_k^i)+e_{\varrho_i}(t_k^i)\big)^2\tilde{\xi}_i^TP^2\tilde{\xi}_i\ge\vartheta_{2i}e^{-\pi_{2i}t_k^i}.\end{aligned}\tag{2.69}$$

Due to the fact that $0\ge\vartheta_{1i}e^{-\pi_{1i}T}>0$ or $0\ge\vartheta_{2i}e^{-\pi_{2i}t_k^i}>0$, which is obviously contradictory. Therefore, the Zeno behavior is excluded under the designed triggering mechanism in (2.57). The proof is completed. □

Subsequently, based on the observer in Eq. (2.55), we design the controller for follower i to stabilize multiagent system

$$u_i=M(x_i-\zeta_i),\ i=1,\dots,N\tag{2.70}$$

where $M=-W^{-1}B^TO$ is designed, and $O>0$ is the solution of Eq. (2.71)

$$A^TO+OA+S-2OBW^{-1}B^TO=0,\tag{2.71}$$

S is given positive definite matrix and W is invertible matrix.

The following theorem shows that the designed controller can stabilize the multiagent systems.

Theorem 2.4 *Under the condition of Assumptions 2.4–2.6, the leader-follower consensus can be reached for the multiagent system* (2.2) *with* $C=I$ *and communication link faults in Eq.* (2.54) *under the adaptive event-triggered observer in Eq.* (2.55) *and the controller in Eq.* (2.70).

Proof Under the controller in (2.70), the following dynamics can be obtained:

$$\dot{\tilde{\epsilon}}_i=(A+BM)\tilde{\epsilon}_i+BM\epsilon_i,\tag{2.72}$$

It follows from (2.71) that matrix $A+BM$ is Hurwitz.

Consider the following Lyapunov function candidate as

$$\tilde{V}_\epsilon = \sum_{i\in I[1,N]} \tilde{\epsilon}_i^T Q_\epsilon \tilde{\epsilon}_i, \tag{2.73}$$

where Q_ϵ is the solution of equation $(A + BM)^T Q_\epsilon + Q_\epsilon(A + BM) = -S_\epsilon$ and S_ϵ is given positive definite matrix. Calculate the derivative of $\tilde{V}_\epsilon$

$$\begin{aligned}\dot{\tilde{V}}_\epsilon =& 2 \sum_{i\in I[1,N]} \tilde{\epsilon}_i^T Q_c\Big((A + BM)^T \epsilon_i + BM\epsilon_i\Big) \\ \leq& - \sum_{i\in I[1,N]} \tilde{\epsilon}_i^T (S_\epsilon - c_\epsilon I)\tilde{\epsilon}_i + \frac{1}{c_\epsilon} \sum_{i\in I[1,N]} \|Q_\epsilon BM\epsilon_i\|^2, \end{aligned}\tag{2.74}$$

where $0 < c_\epsilon < \lambda_{min}(S_\epsilon)$ is given.

Based on Theorem 2.1 and Lemma 1.5, one can obtain that $\lim_{t\to\infty} \epsilon_i = 0$. Therefore, the leader-follower consensus can be reached for the multiagent system with communication link faults. The proof is completed. □

2.5 Simulation Examples

In this section, two examples are introduced to illustrate the effectiveness of the theoretical results.

Example 2.1 Consider the multiagent system in (2.1), the system matrices are given as $A = [0\ 1; -1\ 0]$, $B = [0, 1]$, $C = [1, 0]$. The switching topologies are shown in Fig. 2.1 and the switching time signal satisfies that

$$\sigma(t) = \begin{cases} 1, \ if\ lT_1 \leq t < (l + \frac{1}{3})T_1 \\ 2, \ if\ (l + \frac{1}{3})T_1 \leq t < (l + \frac{2}{3})T_1 \\ 3, \ if\ (l + \frac{2}{3})T_1 \leq t < (l + 1)T_1 \end{cases} \tag{2.75}$$

where $T_1 = 1$s and l denotes the non-negative integer.

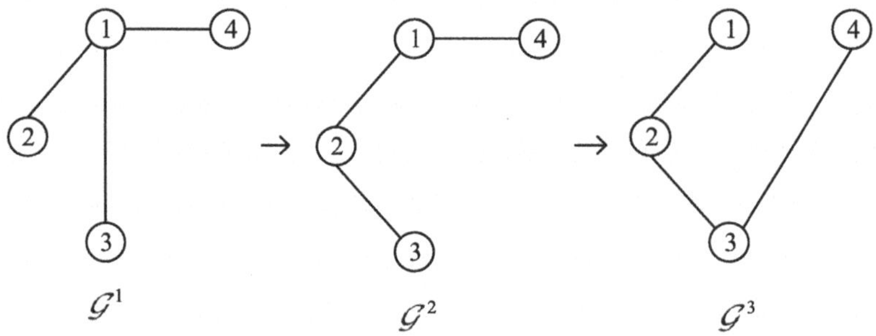

Fig. 2.1 All possible switching topologies, reprinted from ref. Zhang et al. (2022), copyright 2024, with permission from IEEE

By using pole placement method, select matrices F and G such that $A + FC$ and $A + BG$ are Hurwitz matrices. The selection of relevant parameters follows Algorithm 2.1, i.e., $E = [-0.5732, \quad -0.1251]^T$, $k_{ij} = 0.3$, $\iota = 0.9$, $\nu = 5$, $\omega = 0.5$. The initial values of all state variables are randomly selected, and the simulation results under the proposed observer-based event-triggered controller are shown in Figs. 2.2, 2.3, and 2.4.

Figure 2.2 depicts the dynamic response of adaptive coupling weights, from which it can be seen that the final trajectories tend to stabilize, meaning that all adaptive coupling weights converge to some finite value. Figure 2.3 depicts the dynamic response of the consensus error, from which it can be seen that the consensus error eventually tends to zero, indicating that all agents ultimately reach consensus. Figure 2.4 depicts the time instant when the event happens, and it can be observed that the agent achieves intermittent communication without Zeno behavior, thereby achieving the goal of saving communication resources. The simulation results have jointly verified the feasibility of the proposed event-triggering mechanism and controller, demonstrating the solution to the problem of leaderless consensus. Similarly, it is possible to simulate and verify the feasibility of leader-follower consensus and to omit it here.

Example 2.2 Consider a multiagent system composed of six followers and one leader, the dynamics of multiagent system shown in (2.1), where the system matrices are given by

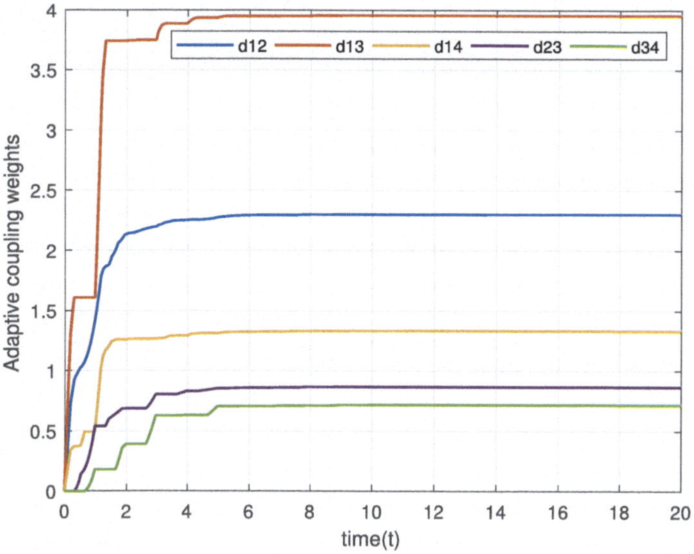

Fig. 2.2 The trajectories of the adaptive coupling weights d_{ij} in (2.4), reprinted from ref. Zhang et al. (2022), copyright 2024, with permission from IEEE

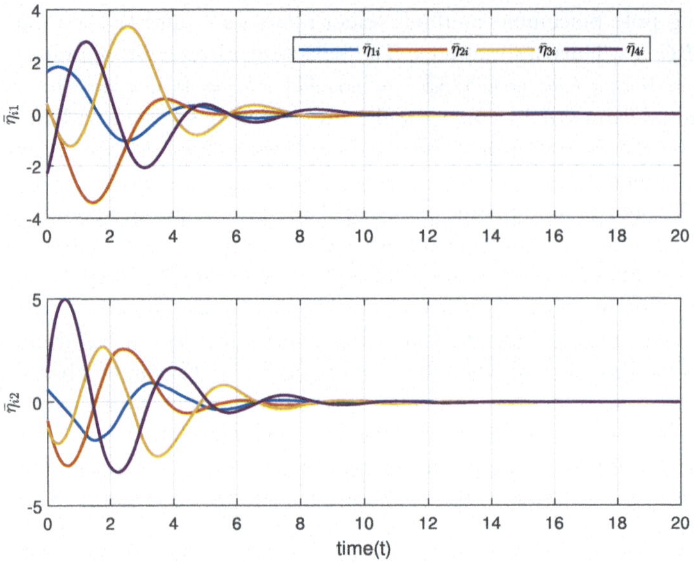

Fig. 2.3 First and second dimensional trajectories of $\bar{\eta}_i$, reprinted from ref. Zhang et al. (2022), copyright 2024, with permission from IEEE

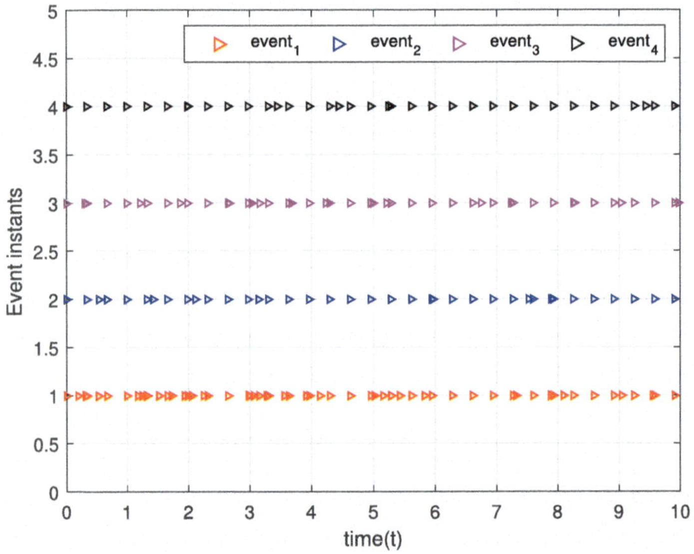

Fig. 2.4 The event instants under triggering mechanism (2.6), reprinted from ref. Zhang et al. (2022), copyright 2024, with permission from IEEE

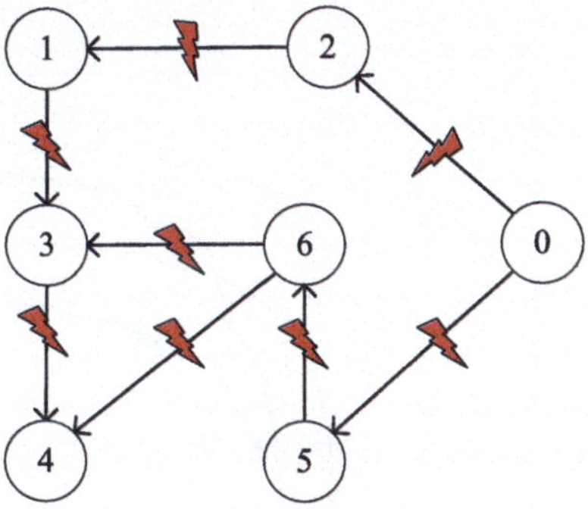

Fig. 2.5 The directed graph $\bar{\mathcal{G}}$ with communication link faults, reprinted from ref. Zhang and Zhang (2022), copyright 2024, with permission from IEEE

$$A = \begin{bmatrix} 0 & 1 & 0 \\ 0 & 0 & 1 \\ -1 & -2 & -1 \end{bmatrix}, \ B = \begin{bmatrix} 0 \\ 0 \\ 1 \end{bmatrix}, \ C = \begin{bmatrix} 1\,0\,0 \\ 0\,1\,0 \\ 0\,0\,1 \end{bmatrix}.$$

The communication way among agents is given in Fig. 2.5, and communication link faults may occur on each edge, where the corrupted weights $\psi_{ij}^{f}(t) = 0.5 * \sin(t) * rand(\cdot)$ and $\psi_{i0}^{f}(t) = 0.5 * \sin(t) * rand(\cdot)$ are set.

In simulation, the initial states $x_i(0)$ of the seven agents and the states $\zeta_i(0)$ of the observers are randomly selected, the relative parameters are given as $\iota_1 = 0.35$, $\iota_2 = 0.35$, $\vartheta_{1i} = 0.5$, $\vartheta_{2i} = 0.5$, $\pi_{1i} = 0.5$, $\pi_{2i} = 0.5$, $i \in I[1, 6]$. Moreover, the initial adaptive weights are set as 1, for $i \in I[1, 6]$. Then, execute the proposed adaptive event-triggered observer in Eq. (2.55) and controller in Eq. (2.70) and obtain simulation results as shown in Figs. 2.6, 2.7, 2.8, and 2.9.

Figure 2.6 shows the dynamic response of the adaptive weights, which implies that adaptive coupling weights converge to some finite values. Figure 2.7 shows the dynamic response of the observer' error for each follower, which indicates that the observer' error tends to 0. Figure 2.8 depicts the inter-event time interval for six followers, which indicates that the time interval between adjacent triggers is greater than zero, so the multiagent system will not exhibit Zeno behavior under the designed triggering mechanism. Figure 2.9 depicts the dynamic response of seven agents, which indicates that the multiagent system ultimately reaches stability under the designed controller. Therefore, the method proposed in this section is feasible for the leader-follower consensus of linear multi-agent systems under directed graphs with communication link faults.

2.6 Summary

This chapter investigated the consensus problem of linear multi-agent systems in the case of switching topologies and communication link faults. Firstly, in the case of

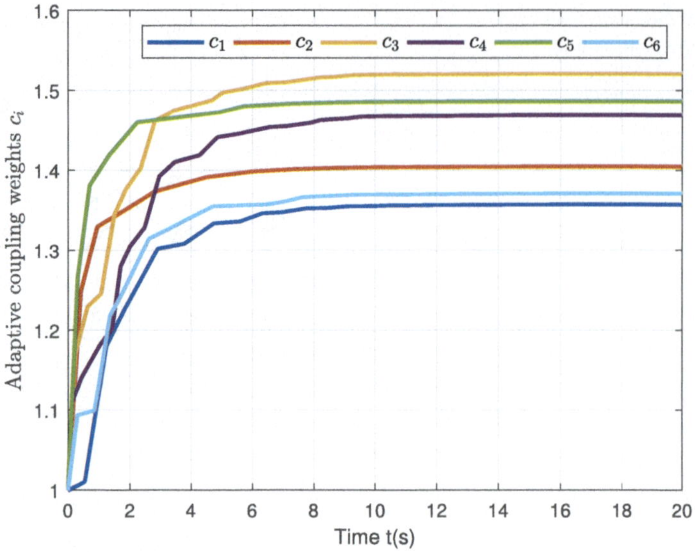

Fig. 2.6 The dynamic response of adaptive weights c_i, reprinted from ref. Zhang and Zhang (2022), copyright 2024, with permission from IEEE

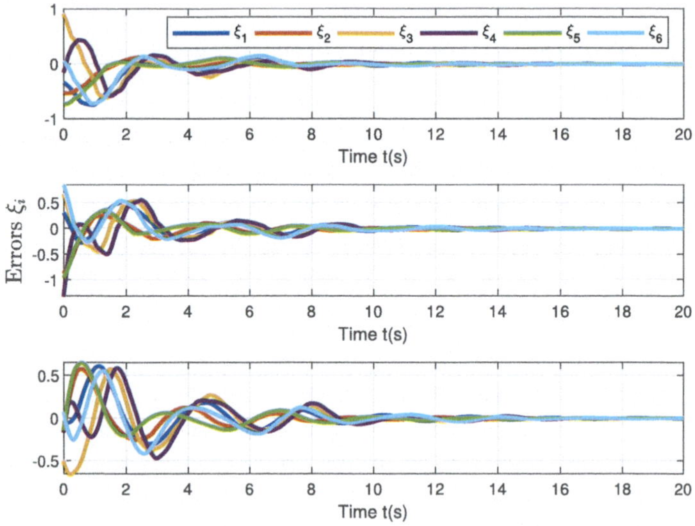

Fig. 2.7 The dynamic response of the observer' error ξ_i, reprinted from ref. Zhang and Zhang (2022), copyright 2024, with permission from IEEE

the switching topologies, the adaptive event-triggered mechanism and the observer-based event-triggered controller were designed to estimate the unmeasurable states of the agents and achieve leaderless consensus and leader-follower consensus. Secondly,

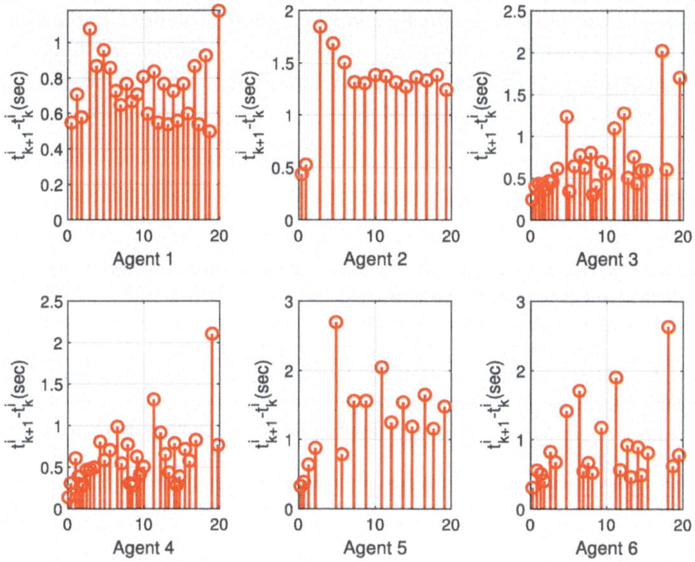

Fig. 2.8 The inter-event times of six followers, reprinted from ref. Zhang and Zhang (2022), copyright 2024, with permission from IEEE

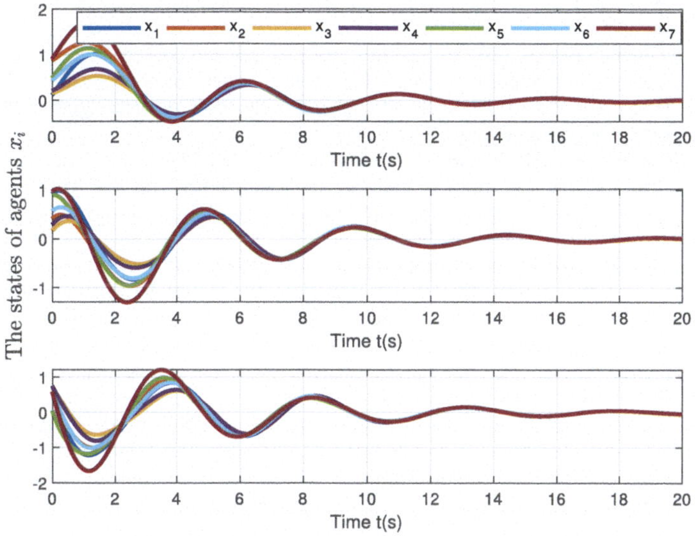

Fig. 2.9 The dynamic response of all agents' states x_i, reprinted from ref. Zhang and Zhang (2022), copyright 2024, with permission from IEEE

in the case of communication link faults, the adaptive event-triggered observer was designed to estimate the leader's information under communication link faults. Based on the proposed observer, the controller was designed to achieve the leader-follower consensus problem under communication link faults. Finally, the feasibility of the proposed methods was verified through theoretical analysis and simulation.

References

Adaldo A, Liuzza D, Dimarogonas D, JOhansson K (2015) Control of multi-agent systems with event-triggered cloud access. ECC, pp 1–6

Chen C, Xie K, Lewis F, Xie S, Fierro R (2020) Adaptive synchronization of multi-agent systems with resilience to communication link faults. Automatica 111:108636

Chen T, Liu X, Lu W (2007) Pinning complex networks by a single controller. IEEE Trans Circuits Syst I: Regular Papers 54(6):1317–1326

Hu J, Hong Y (2007) Leader-following coordination of multi-agent systems with coupling time delays. Physica A 374:853–863

Khalil HK (2012) Nonlinear systems, 3rd edn. Prentice Hall, Upper Saddle River

Krstić M, Kanellakopoulos I, Kokotovic P (1995) Nonlinear and adaptive control design. Wiley, New York

Li Z, Duan Z, Chen G, Huang L (2010a) Consensus of multiagent systems and synchronization of complex networks: a unified viewpoint. IEEE Trans Circuits Syst I: Regular Papers 57(1):213–224

Li Z, Duan Z, Chen G, Huang L (2010b) Consensus of multiagent systems and synchronization of complex networks: a unified viewpoint. IEEE Trans Circuits Syst I: Regular Papers 57(1)

Li Z, Ren W, Liu X, Xie L (2013) Distributed consensus of linear multi-agent systems with adaptive dynamic protocols. Automatica 49:1986–1995

Li Z, Wen G, Duan Z, Ren W (2015) Designing fully distributed consensus protocols for linear multi-agent systems with directed graphs. IEEE Trans Autom Control 60(4):1152–1157

Liu K, Ji Z, Zhang X (2020) Periodic event-triggered consensus of multi-agent systems under directed topology. Neurocomputing 385(14)

Liuzza D, Dimarogonas D, Bernardo M, Johansson K (2013) Distributed model based event-triggered control for synchronization of multi-agent systems. IFAC, pp 329–334

Lu W, Chen T (2006) New approach to synchronization analysis of linearly coupled ordinary differential systems. Physica D 213:214–230

Lu W, Han Y, Chen T (2015) Synchronization in networks of linearly coupled dynamical systems via event-triggered diffusions. IEEE Trans Neural Netw Learn Syst 26(12):3060–3069

Meng X, Chen T (2013) Event based agreement protocols for multi-agent networks. Automatica 49(7)

Ni W, Cheng D (2010) Leader-following consensus of multi-agent systems under fixed and switching topologies. Syst Control Lett 59:209–217

Wen G, Hu G, Yu W, Chen G (2014) Distributed $\mathcal{H}_\infty$ consensus of higher order multiagent systems with switching topologies. IEEE Trans Circuits Syst II: Express Briefs 61(5):359–363

Yi X, Lu W, Chen T (2017) Pull-based distributed event-triggered consensus for multiagent systems with directed topologies. IEEE Trans Neural Netw Learn Syst 28(1):71–79

Zhang H, Han J, Wang Y, Jiang H (2019) h_∞ consensus for linear heterogeneous multiagent systems based on event-triggered output feedback control scheme. IEEE Trans Cybern 49(6):2268–2279

Zhang J, Dai J, Zhang H, Sun S (2020) Cooperative output regulation of heterogeneous linear multi-agent systems based on the event-triggered distributed control under switching topologies. Appl Math Comput 390:125611

Zhang H, Duan J, Wang Y, Gao Z (2021) Bipartite fixed-time output consensus of heterogeneous linear multiagent systems. IEEE Trans Cybern 51(2):548–557

Zhang J, Zhang H, Zhang K, Cai Y (2022) Observer-based output feedback event-triggered adaptive control for linear multiagent systems under switching topologies. IEEE Trans Neural Netw Learn Syst 33(12):7161–7171

Zhang H, Feng G, Yan H, Chen Q (2013) Observer-based output feedback event-triggered control for consensus of multi-agent systems. IEEE Trans Ind Electron 61(9):4885–4894

Zhang J, Zhang H (2022) Adaptive event-triggered consensus of linear multiagent systems with resilience to communication link faults for digraphs. IEEE Trans Circuits Syst II: Express Briefs 69(7):3249–3253

Chapter 3
Consensus Control of Nonlinear Multiagent Systems Under Asynchronous Communication

3.1 Introduction

The adaptive distributed observer-based event-triggered consensus protocol proposed in Chap. 2 solves the consensus problem of linear multiagent systems. The designed event-triggered mechanism not only ensures that the agent measures relevant information at discrete times, but also achieves intermittent communication between the agents. However, the proposed event-triggered mechanism is designed based on node. When the triggering condition of agent i is satisfied, agent i will simultaneously transmit information to all its neighbors.

To avoid the shortcomings of node-based event-triggered mechanisms, the novel edge-based event-triggered mechanisms are proposed in Cao et al. (2016), Xiao et al. (2015), Cheng et al. (2021), Meng et al. (2018), and Cheng and Li (2019b). It can be observed that under the edge-based event-triggered mechanism, the agent achieves intermittent asynchronous communication between corresponding neighbors. Moreover, the construction of triggering functions in the edge-based event-triggered mechanisms only relies on the information of the two agents that make up the edge, so its design is simple. The authors in Meng et al. (2018) prove that compared to node-based event-triggered mechanism, the edge-based event-triggered mechanism can save system communication resources. The authors in Cheng et al. (2021) propose the edge-based event-triggered control protocol to solve the formation problem of linear multiagent systems. In this case, intermittent asynchronous communication is realized between the neighbors, i.e., when the edge-based event-triggered mechanism is satisfied, only the two agents constituting the edge transmit the information to each other. The authors in Cheng and Li (2019b) design the edge-based event-triggered control strategy to solve the leader-follower consensus problem in linear multiagent systems with non-zero input leaders. It is worth mentioning that the method in Cheng and Li (2019b) ensures asynchronous communication between neighbors, that is, when the edge-based event-triggered mechanism is satisfied, the agent i only transmits information to the agent j instead of all its out-neighbors, and the agent

J. Zhang et al., *Adaptive Collaborative Control of Multiagent Systems by Event-Triggered Mechanisms*, Intelligent Control and Learning Systems 14,
https://doi.org/10.1007/978-981-97-7866-9_3

j may not necessarily transmit information to the agent i. In contrast to the edge-based event-triggered mechanism, the node-based event-triggered mechanism needs to design a small number of triggering mechanisms, which is equal to the number of nodes. The number of triggering mechanisms designed by the edge-based event-triggered mechanism is equal to the number of edges. Obviously, the number of edges is greater than the number of nodes; for this reason, the edge-based event-triggered mechanism requires high system hardware conditions and should be weighed in the actual use of the choice of which type of triggering mechanism to design.

Due to the system itself or external influences, the actual system has certain nonlinear characteristics, and in most cases only some parts with small influences can be neglected. Therefore, it is very necessary to study the control problem for nonlinear systems (Rehan et al. 2019; Chu et al. 2016; Ahmad and Rehan 2016). Inspired by the above articles, this paper investigates the consensus problems of nonlinear multiagent systems with one-sided Lipschitz conditions and nonlinear multiagent systems with Lipschitz conditions by designing the novel edge-based event-triggered mechanisms and node-based event-triggered mechanisms.

3.2 Problem Formulation

Consider the multiagent system composed of N followers and one leader, the dynamics of ith follower and leader are modeled by

$$\dot{x}_i(t) = \mathcal{A}x_i(t) + \mathcal{B}u_i(t) + f(x_i, t), \ \forall\, i \in I[1, N], \tag{3.1}$$

and

$$\dot{x}_0(t) = \mathcal{A}x_0(t) + f(x_0, t), \tag{3.2}$$

where $x_i \in R^n$, $u_i \in R^m$ denotes the state and input of follower i, respectively, $f(\cdot) : R^n \times R \rightarrow R^n$ denotes the uncertain nonlinear function, and $x_0 \in R^n$ denotes the state of the leader 0.

The control objective of this chapter is to achieve the leader-follower consensus problem of nonlinear multiagent system composed of (3.1) and (3.2). To achieve the leader-follower consensus, the following assumptions are given about matrices $\mathcal{A}$, $\mathcal{B}$, the nonlinear term $f(\cdot)$, and communication topology graph $\bar{\mathcal{G}}$.

Assumption 3.1 The nonlinear function $f(\cdot)$ satisfies the Lipschitz condition. That is, there exists a positive constant α such that the following condition satisfies

$$\|f(x_i(t), t) - f(x_j(t), t)\| \leq \alpha\|x_i(t) - x_j(t)\|, \ i, j \in I[1, \ldots, N]. \tag{3.3}$$

Assumption 3.2 The pair $(\mathcal{A}, \mathcal{B})$ is stabilizable, the subgraph $\mathcal{G}$ among followers is undirected and connected, and the graph $\bar{\mathcal{G}}$ contains a directed spanning tree with, and root node is the leader.

Assumption 3.3 There exist $\varrho \in R, x_i \in R^n, x_j \in R^n$ such that nonlinear function $f(\cdot) \in R^p$ satisfies the following one-sided Lipschitz condition:

$$\Big(f(x_i) - f(x_j)\Big)'(x_i - x_j) \leq \varrho \|x_i - x_j\|^2, \tag{3.4}$$

and ϱ is called as the one-sided Lipschitz constant.

Assumption 3.4 There exist $\sigma \in R, \varsigma \in R, x_i \in R^n, x_j \in R^n$ such that nonlinear function $f(\cdot) \in R^p$ satisfies the quadratic inner boundedness condition,

$$\|f(x_i) - f(x_j)\|^2 \leq \sigma \|x_i - x_j\|^2 + \varsigma (x_i - x_j)'\Big(f(x_i) - f(x_j)\Big), \tag{3.5}$$

and σ, ς are called as the quadratic inner boundedness constants.

Remark 3.1 Assumptions 3.1 and 3.2 are standard in dealing with the cooperative control problems for multiagent system. Assumption 3.1 generally appears in nonlinear multiagent system. Assumption 3.1 always can be found in dealing with the leader-follower consensus problem.

3.3 Consensus Control for Multiagent Systems with Lipschitz Dynamics by the Edge-Based Event-Triggered Mechanisms

This section is divided into two subsections, the design of event-triggered mechanism and consensus controller, and the theoretical analysis of the effectiveness of the proposed methods.

3.3.1 Node-Based Event-Triggered Mechanisms and Edge-Based Event-Triggered Mechanisms

Most existing results assume that leader can continuously broadcast information to informed followers, which requires agents acting as leaders to have ideal communication resources. In fact, each agent is usually equipped with limited communication resources, so it is crucial to avoid continuous transmission of information between all agents. Therefore, this subsection proposes three kinds of triggering mechanisms to further achieve asynchronous transmission of information from leaders to followers and followers to followers.

In this subsection, agents are divided into leader, informed followers, and uninformed followers. Record the instants when the leader transmits information to all informed followers as t_k^0, the instants when the leader transmits information by the edge $(i, 0)$ to informed follower i as t_k^{i0}, and the instants when the follower j transmits information to the follower i through the edge (i, j) as t_k^{ij}. The triggering instants t_k^0, t_k^{i0}, t_k^{ij} depend on the following triggering mechanisms:

$$\begin{aligned} t_k^0 &= \inf\{t > t_{k-1}^0 | h_0(t) \geq 0\} \\ t_k^{i0} &= \inf\{t > t_{k-1}^0 | h_{i0}(t) \geq 0\} \\ t_k^{ij} &= \inf\{t > t_{k-1}^0 | h_{ij}(t) \geq 0\} \end{aligned} \tag{3.6}$$

where $h_0(t)$ denotes the triggering function for the leader, and its construction depends on the leader's information, h_{i0} denotes the triggering function for the edge $(i, 0)$, and its construction depends on relative information for the leader 0 and follower i, h_{ij} denotes the triggering function for the edge (i, j), and its construction depends on relative information for the followers i and j. Note that the first one in Eq. (3.6) is called as the node-based event-triggered mechanism and the second and third ones in Eq. (3.7) are called as the edge-based event-triggered mechanisms.

Define the measurement errors $e_0(t)$, $e_{i0}(t)$, $e_{ij}(t)$ as

$$\begin{aligned} e_0(t) &= \tilde{x}_0(t) - x_0(t), \ \ t \in [t_k^0, t_{k+1}^0), \\ e_{i0}(t) &= \tilde{x}_{i0}(t) - x_i(t), \ \ t \in [t_k^{i0}, t_{k+1}^{i0}), \\ e_{ij}(t) &= \tilde{x}_{ij}(t) - x_i(t), \ \ t \in [t_k^{ij}, t_{k+1}^{ij}), \end{aligned} \tag{3.7}$$

where $\tilde{x}_0(t)$, $\tilde{x}_{i0}(t)$, $\tilde{x}_{ij}(t)$ denote the open-loop estimation of $x_0(t)$, $x_{i0}(t)$, $x_{ij}(t)$, and satisfy

$$\begin{aligned} \tilde{x}_0(t) &= e^{\mathcal{A}(t-t_k^0)} x_0(t_k^0), \ \ t \in [t_k^0, t_{k+1}^0), \\ \tilde{x}_{i0}(t) &= e^{\mathcal{A}(t-t_k^{i0})} x_i(t_k^{i0}), \ \ t \in [t_k^{i0}, t_{k+1}^{i0}), \\ \tilde{x}_{ij}(t) &= e^{\mathcal{A}(t-t_k^{ij})} x_i(t_k^{ij}), \ \ t \in [t_k^{ij}, t_{k+1}^{ij}). \end{aligned} \tag{3.8}$$

Remark 3.2 Compared with references in Li et al. (2013), Cheng and Li (2019a), the proposed node-based event-triggered mechanism can effectively avoid leader transmitting information continuously to all followers and can effectively reduce the consumption of leader communication resources. Compared with references in Zhang et al. (2020a), Yang et al. (2019), the designed edge-based event-triggered mechanisms can achieve asynchronous event-triggered communication between leaders and informed followers, which can further achieve efficient information transmission.

3.3.2 Static Edge-Based Event-Triggered Consensus Protocol

Design the static edge-based event-triggered consensus control protocol for follower i

$$u_i = d\mathcal{F}a_{i0}(\tilde{x}_{i0} - \tilde{x}_0) + d\mathcal{F}\sum_{i\in I[1,N]} a_{ij}(\tilde{x}_{ij} - \tilde{x}_{ji}) \tag{3.9}$$

where $\mathcal{F} \in R^{m\times n}$ is the feedback matrix to be designed and $d > 0$ is a positive constant.

Construct the measurement errors as follows:

$$\begin{aligned} e_0 &= \tilde{x}_0 - x_0 \\ e_{ij} &= \tilde{x}_{ij} - x_i. \end{aligned} \tag{3.10}$$

Then, design the triggering functions in Eq. (3.6) as follows:

$$\begin{aligned} h_0(t) =& -\nu_0 e^{-\omega_0 t} + e_0^T \mathcal{M} e_0 \\ h_{i0}(t) =& -\frac{1}{2}(\tilde{x}_{i0} - \tilde{x}_0)^T \mathcal{M}(\tilde{x}_{i0} - \tilde{x}_0) - \nu_{i0} e^{-\omega_{i0} t} + e_{i0}^T \mathcal{M} e_{i0} \\ h_{ij}(t) =& -\frac{1}{4}(\tilde{x}_{ij} - \tilde{x}_{ji})^T \mathcal{M}(\tilde{x}_{ij} - \tilde{x}_{ji}) - \nu_{ij} e^{-\omega_{ij} t} + e_{ij}^T \mathcal{M} e_{ij} \end{aligned} \tag{3.11}$$

where $\mathcal{M} \in R^{n\times n}$ is a feedback matrix to be designed and $\nu_0,\ \nu_{i0},\ \nu_{ij}, \omega_0, \omega_{i0},\ \omega_{ij}$ are positive constants to be designed, for $i, j \in I[1, N]$.

Remark 3.3 According to the trigger function in (3.12), the node-based event-triggered function only depends on the leader's information, while edge-based event-triggered function depends on their corresponding edges. The advantage of designed triggering function is to achieve asynchronous event-triggered communication between agents and save communication resources, and its construction is simple. Specifically, based on the node-based event-triggered mechanism, when the event occurs for agent i, if there exists neighbor j of agent i, whose state information is similar, agent i still transmits information to agent j, which obviously causes communication resource waste. The edge-based event-triggered mechanism compensates for this deficiency.

Construct the consensus error ζ_i as follows:

$$\zeta_i = x_i - x_0, \tag{3.12}$$

and its compact form is $\zeta = \text{col}(\zeta_1, \ldots, \zeta_N)$. Then, calculate the derivative of ζ_i by Eqs. (3.1) and (3.2)

$$\dot{\zeta}_i = \mathcal{A}\zeta_i + d\mathcal{B}\mathcal{F}\sum_{j\in I[1,N]} a_{ij}(\tilde{x}_{ij} - \tilde{x}_{ji}) + d\mathcal{B}\mathcal{F}a_{i0}(\tilde{x}_{i0} - \tilde{x}_0) + f(x_i,t) - f(x_0,t). \tag{3.13}$$

Before presenting the main results, provide an algorithm to give the relative parameters for the triggering function and the controller.

Algorithm 3.1 Under Assumptions 3.1 and 3.2, the related parameters of the distributed controller (3.9) and triggering functions (3.11) are obtained according to the following steps:

1. Choose the feedback matrices as $\mathcal{F} = -\mathcal{B}^T Q^{-1}$, $\mathcal{M} = Q^{-1}\mathcal{B}\mathcal{B}^T Q^{-1}$. $Q > 0$ is a positive definite matrix satisfying the following linear matrix inequality (LMI):

$$\begin{bmatrix} \mathcal{A}Q + Q\mathcal{A}^T - \varrho\mathcal{B}\mathcal{B}^T + \alpha^2 I & Q \\ Q & -I \end{bmatrix} < 0, \tag{3.14}$$

where $\varrho > 0$ is a positive constant and $\alpha > 0$ is a positive Lipschitz constant.
2. Choose parameters $\nu_0,\ \nu_{i0},\ \nu_{ij},\ \omega_0,\ \omega_{i0},\ \omega_{ij}$ to be positive constants, for $i,\ j \in I[1,N]$, and $d \geq \frac{\varrho}{\lambda_1}$, where λ_1 is the smallest eigenvalue of the matrix $\mathcal{H}$.

Remark 3.4 In Algorithm 3.1, the matrices $\mathcal{F}$ and $\mathcal{M}$ depend on the solution of LMI in Eq. (3.14). In Li et al. (2013), it is explained that under the conditions of Algorithm 3.1, LMI in Eq. (3.14) must have solution.

Theorem 3.1 *Under the condition of Assumptions 3.1 and 3.2, the static edge-based event-triggered consensus control protocol in Eq.* (3.9) *and triggering functions in Eq.* (3.11) *are constructed through Algorithm 3.1, then the leader-follower consensus problem of multiagent system of Eqs.* (3.1) *and* (3.2) *is solved.*

Proof Consider the following Lyapunov function $\mathcal{V}_1$,

$$\mathcal{V}_1 = \frac{1}{2}\sum_{i\in I[1,N]} \zeta_i^T Q^{-1}\zeta_i \tag{3.15}$$

where Q is given in Algorithm 3.1. Calculate the derivative of $\mathcal{V}_1$ through Eq. (3.13) as

$$\begin{aligned} \dot{\mathcal{V}}_1 = &\sum_{i\in I[1,N]} \zeta_i^T Q^{-1}\mathcal{A}\zeta_i - d\sum_{i\in I[1,N]}\sum_{j\in I[1,N]} a_{ij}\zeta_i^T\mathcal{M}(\tilde{x}_{ij} - \tilde{x}_{ji}) \\ &- d\sum_{i\in I[1,N]} a_{i0}\zeta_i^T\mathcal{M}(\tilde{x}_{i0} - \tilde{x}_0) + \sum_{i\in I[1,N]} \zeta_i^T Q^{-1}\big(f(x_i,t) - f(x_0,t)\big). \end{aligned} \tag{3.16}$$

According to the definition of consensus error in Eq. (3.12) and measurement errors in Eq. (3.10), and Lemma 1.6, one can obtain that

$$
\begin{aligned}
& - d \sum_{i\in I[1,N]} \sum_{j\in I[1,N]} a_{ij}\zeta_i^T \mathcal{M}(\tilde{x}_{ij} - \tilde{x}_{ji}) \\
= & -\frac{d}{4} \sum_{i\in I[1,N]} \sum_{j\in I[1,N]} a_{ij}(\tilde{x}_{ij} - \tilde{x}_{ji})^T \mathcal{M}(\tilde{x}_{ij} - \tilde{x}_{ji}) \\
& -\frac{d}{4} \sum_{i\in I[1,N]} \sum_{j\in I[1,N]} a_{ij}(\zeta_i - \zeta_j)^T \mathcal{M}(\zeta_i - \zeta_j) \\
& +\frac{d}{4} \sum_{i\in I[1,N]} \sum_{j\in I[1,N]} a_{ij}(e_{ij} - e_{ji})^T \mathcal{M}(e_{ij} - e_{ji}) \\
\leq & -\frac{d}{4} \sum_{i\in I[1,N]} \sum_{j\in I[1,N]} a_{ij}(\tilde{x}_{ij} - \tilde{x}_{ji})^T \mathcal{M}(\tilde{x}_{ij} - \tilde{x}_{ji}) \\
& -\frac{d}{4} \sum_{i\in I[1,N]} \sum_{j\in I[1,N]} a_{ij}(\zeta_i - \zeta_j)^T \mathcal{M}(\zeta_i - \zeta_j) \\
& + d \sum_{i\in I[1,N]} \sum_{j\in I[1,N]} a_{ij} e_{ij}^T \mathcal{M} e_{ij}
\end{aligned}
\tag{3.17}
$$

and

$$
\begin{aligned}
& - d \sum_{i\in I[1,N]} a_{i0}\zeta_i^T \mathcal{M}(\tilde{x}_{i0} - \tilde{x}_0) \\
= & -\frac{d}{2} \sum_{i\in I[1,N]} a_{i0}(\tilde{x}_{i0} - \tilde{x}_0)^T \mathcal{M}(\tilde{x}_{i0} - \tilde{x}_0) \\
& +\frac{d}{2} \sum_{i\in I[1,N]} a_{i0}(e_{i0} - e_0)^T \mathcal{M}(e_{i0} - e_0) \\
& -\frac{d}{2} \sum_{i\in I[1,N]} a_{i0}\zeta_i^T \mathcal{M}\zeta_i \\
\leq & -\frac{d}{2} \sum_{i\in I[1,N]} a_{i0}\zeta_i^T \mathcal{M}\zeta_i - \frac{d}{2} \sum_{i\in I[1,N]} a_{i0}(\tilde{x}_{i0} - \tilde{x}_0)^T \mathcal{M}(\tilde{x}_{i0} - \tilde{x}_0) \\
& + d \sum_{i\in I[1,N]} a_{i0} e_{i0}^T \mathcal{M} e_{i0} + d \sum_{i\in I[1,N]} a_{i0} e_0^T \mathcal{M} e_0.
\end{aligned}
\tag{3.18}
$$

Taking advantage of Assumption 3.1, we have

$$\begin{aligned}
&\sum_{i\in I[1,N]} \zeta_i^T Q^{-1}\big(f(x_i,t)-f(x_0,t)\big)\\
&\quad\le \alpha \sum_{i\in I[1,N]} \|\zeta_i^T Q^{-1}\|\|\zeta_i\|\\
&\quad\le \frac{1}{2}\sum_{i\in I[1,N]} \zeta_i^T(\alpha^2 Q^{-2}+I_n)\zeta_i.
\end{aligned} \tag{3.19}$$

It follows from Eqs. (3.17)–(3.19) that Eq.(3.16) satisfies

$$\begin{aligned}
\dot{\mathcal{V}}_1 \le& \frac{1}{2}\sum_{i\in I[1,N]} \zeta_i^T(Q^{-1}\mathcal{A}+\mathcal{A}^TQ^{-1}+\alpha^2Q^{-2}+I_n)\zeta_i\\
&-\frac{d}{2}\sum_{i\in I[1,N]}\sum_{j\in I[0,N]} a_{ij}\zeta_i^T\mathcal{M}(\zeta_i-\zeta_j)\\
&+d\sum_{i\in I[1,N]}\sum_{j\in I[1,N]} a_{ij}\big(e_{ij}^T\mathcal{M}e_{ij}-\frac{1}{4}(\tilde{x}_{ij}-\tilde{x}_{ji})^T\mathcal{M}(\tilde{x}_{ij}-\tilde{x}_{ji})\big)\\
&+d\sum_{i\in I[1,N]} a_{i0}\big(e_{i0}^T\mathcal{M}e_{i0}-\frac{1}{2}(\tilde{x}_{i0}-\tilde{x}_0)^T\mathcal{M}(\tilde{x}_{i0}-\tilde{x}_0)\big)\\
&+d\sum_{i\in I[1,N]} a_{i0}e_0^T\mathcal{M}e_0.
\end{aligned} \tag{3.20}$$

Simplify Eq. (3.20) into Eq. (3.21) using triggering mechanism composed of Eqs. (3.6) and (3.11),

$$\begin{aligned}
\dot{\mathcal{V}}_1 \le& \frac{1}{2}\sum_{i\in I[1,N]} \zeta_i^T(Q^{-1}\mathcal{A}+\mathcal{A}^TQ^{-1}+\alpha^2Q^{-2}+I_n)\zeta_i\\
&-\frac{d}{2}\sum_{i\in I[1,N]}\sum_{j\in I[0,N]} a_{ij}\zeta_i^T\mathcal{M}(\zeta_i-\zeta_j)+\varpi
\end{aligned} \tag{3.21}$$

where $\varpi = d\sum_{i\in I[1,N]}(\sum_{j\in I[1,N]} a_{ij}\nu_{ij}e^{\omega_{ij}t}+a_{i0}\nu_0 e^{\omega_0 t})$ is given.

To further simplify, let $\hat{\zeta}_i = Q^{-1}\zeta_i$, $\hat{\zeta} = \mathrm{col}(\hat{\zeta}_1,\ldots,\hat{\zeta}_N)$. Then, we obtain

$$\begin{aligned}
\dot{\mathcal{V}}_1 \le& \frac{1}{2}\sum_{i\in I[1,N]} \hat{\zeta}_i^T(\mathcal{A}Q+Q\mathcal{A}^T+\alpha^2I_n+Q^2)\hat{\zeta}_i\\
&-\frac{d}{2}\sum_{i\in I[1,N]}\sum_{j\in I[0,N]} a_{ij}\hat{\zeta}_i^T\mathcal{B}\mathcal{B}^T(\hat{\zeta}_i-\hat{\zeta}_j)+\varpi\\
=&\frac{1}{2}\hat{\zeta}^T\big(I_N\otimes(\mathcal{A}Q+Q\mathcal{A}^T+\alpha^2I_n+Q^2)-d(\mathcal{H}\otimes\mathcal{B}\mathcal{B}^T)\big)\hat{\zeta}+\varpi
\end{aligned} \tag{3.22}$$

It follows from Lemma 1.2 that $\mathcal{H}$ is a positive definite matrix. Thus, there exists a diagonal block matrix Γ such that $\Gamma^T \mathcal{H} \Gamma = \text{diag}\{\lambda_1, \lambda_2 \cdots, \lambda_N\} = \Lambda$. Let $\tilde{\zeta} = (\Gamma^T \otimes I_n)\hat{\zeta} = \text{col}(\tilde{\zeta}_1, \ldots, \tilde{\zeta}_N)$. Then, Eq. (3.22) satisfies

$$\begin{aligned} \dot{\mathcal{V}}_1 \leq & \frac{1}{2}\tilde{\zeta}^T \left(I_N \otimes (\mathcal{A}\mathcal{Q} + \mathcal{Q}\mathcal{A}^T + \alpha^2 I_n + \mathcal{Q}^2) - d(\Lambda \otimes \mathcal{B}\mathcal{B}^T)\right)\tilde{\zeta} + \varpi \\ = & \frac{1}{2} \sum_{i \in I[1,N]} \tilde{\zeta}_i (\mathcal{A}\mathcal{Q} + \mathcal{Q}\mathcal{A}^T + \alpha^2 I_n + \mathcal{Q}^2 - d\lambda_i \mathcal{B}\mathcal{B}^T)\tilde{\zeta}_i + \varpi \\ \leq & \frac{1}{2} \sum_{i \in I[1,N]} \tilde{\zeta}_i (\mathcal{A}\mathcal{Q} + \mathcal{Q}\mathcal{A}^T + \alpha^2 I_n + \mathcal{Q}^2 - d\lambda_1 \mathcal{B}\mathcal{B}^T)\tilde{\zeta}_i + \varpi \\ = & \frac{1}{2} \sum_{i \in I[1,N]} \tilde{\zeta}_i \Omega \tilde{\zeta}_i + \varpi \end{aligned} \tag{3.23}$$

where $\Omega = \mathcal{A}\mathcal{Q} + \mathcal{Q}\mathcal{A}^T + \alpha^2 I_n + \mathcal{Q}^2 - d\lambda_1 \mathcal{B}\mathcal{B}^T$ is given. Since $d \geq \frac{\varrho}{\lambda_1}$, based on Lemma 1.8 and Algorithm 3.1, one obtains

$$\Omega \leq \mathcal{A}\mathcal{Q} + \mathcal{Q}\mathcal{A}^T + \alpha^2 I_n + \mathcal{Q}^2 - \varrho \mathcal{B}\mathcal{B}^T < 0. \tag{3.24}$$

Thus, there exists a a positive definite matrix $\mathcal{P}$ such that $\Omega \leq -\mathcal{P} < 0$ and

$$\dot{\mathcal{V}}_1 \leq -\frac{1}{2} \sum_{i \in I[1,N]} \tilde{\zeta}_i^T \mathcal{P} \tilde{\zeta}_i + \varpi \leq -\frac{\lambda_{min}(\mathcal{P})\lambda_{min}(\mathcal{Q}^{-2})}{2}\zeta^T \zeta + \varpi. \tag{3.25}$$

Set $\theta \leq \frac{\lambda_{min}(\mathcal{P})\lambda_{min}(\mathcal{Q}^{-2})}{\lambda_{max}(\mathcal{Q}^{-1})}$. Then Eq. (3.25) can be written as

$$\dot{\mathcal{V}}_1 \leq -\theta \mathcal{V}_1 + \varpi. \tag{3.26}$$

According to the Comparison lemma (Khalil 2002), one has

$$\begin{aligned} \mathcal{V}_1(t) \leq & \mathcal{V}_1(0)e^{-\theta t} + d \sum_{i \in I[1,N]} \Big(\sum_{j \in I[0,N]} a_{ij} \nu_{ij} \varphi(\theta, \omega_{ij}, t) \\ & + a_{i0} \nu_0 \varphi(\theta, \omega_0, t) \Big) \end{aligned}$$

where

$$\varphi(\theta, \omega, t) = \begin{cases} te^{-\theta t}, & if \ \theta = \omega \\ \frac{1}{\omega - \theta}(e^{-\theta t} - e^{-\omega t}), & otherwise \end{cases} \tag{3.27}$$

is given. Obviously, $\lim_{t\to\infty} \varphi(\theta, \omega, t) = 0$, which indicates that $\lim_{t\to\infty} \mathcal{V}_1(t) \leq 0$. Combining $\mathcal{V}_1(t) > 0$ and $\lim_{t\to\infty} \mathcal{V}_1(t) \leq 0$, we can obtain that $\lim_{t\to\infty} \mathcal{V}_1(t) = 0$. Therefore, one can obtain that $\lim_{t\to\infty} \zeta(t) = 0$, and the leader-follower consensus for multiagent system composed of (3.1) and (3.2) is reached. That is the end of proof.

3.3.3 Dynamic Edge-Based Event-Triggered Consensus Protocol

The previous section proposes the static edge-based event-triggered consensus protocol, whose design relies on the global information of the communication topology and does not have scalability. To avoid this deficiency, this section proposes the dynamic edge-based event-triggered consensus protocol.

Design the dynamic edge-based event-triggered consensus protocol for follower i,

$$\begin{aligned} u_i &= \mathcal{F} \sum_{j \in I[1,N]} d_{ij} a_{ij} (\tilde{x}_{ij} - \tilde{x}_{ji}) + \mathcal{F} d_{i0} a_{i0} (\tilde{x}_{i0} - \tilde{x}_0) \\ \dot{d}_{ij} &= c_{ij} a_{ij} (\tilde{x}_{ij} - \tilde{x}_{ji})^T \mathcal{M} (\tilde{x}_{ij} - \tilde{x}_{ji}) \\ \dot{d}_{i0} &= c_{i0} a_{i0} (\tilde{x}_{i0} - \tilde{x}_0)^T \mathcal{M} (\tilde{x}_{i0} - \tilde{x}_0) \end{aligned} \tag{3.28}$$

where d_{ij} $\left(d_{ij}(0) = d_{ji}(0)\right)$ is the adaptive coupling weight, $c_{ij} = c_{ji}$ is the positive constant, and $\mathcal{F}$, $\mathcal{M}$ are the feedback gain matrices constructed in Algorithm 3.1.

If not emphasized, the definition of variables is the same as in the previous section. The triggering mechanism is given in Eq. (3.6), and the triggering functions are designed as

$$\begin{aligned} h_0(t) &= e_0^T \mathcal{M} e_0 - \nu_0 e^{-\omega_0 t} \\ h_{i0}(t) &= (1 + 2\iota d_{i0}) e_{i0}^T \mathcal{M} e_{i0} - \nu_{i0} e^{-\omega_{i0} t} - \frac{1}{2} (\tilde{x}_{i0} - \tilde{x}_0)^T \mathcal{M} (\tilde{x}_{i0} - \tilde{x}_0) \\ h_{ij}(t) &= (1 + 2\iota d_{ij}) e_{ij}^T \mathcal{M} e_{ij} - \nu_{ij} e^{-\omega_{ij} t} - \frac{1}{4} (\tilde{x}_{ij} - \tilde{x}_{ji})^T \mathcal{M} (\tilde{x}_{ij} - \tilde{x}_{ji}) \end{aligned} \tag{3.29}$$

where ι, ν_0, ν_{i0}, ν_{ij}, ω_0, ω_{i0}, ω_{ij} $i, j \in I[1, N]$ are positive known constants.

In this case, calculate the derivative of ζ_i based on the adaptive controller in Eq. (3.28) as

$$\begin{aligned} \dot{\zeta}_i &= \mathcal{A} \zeta_i + \mathcal{B} \mathcal{F} \sum_{j \in I[1,N]} d_{ij} a_{ij} (\tilde{x}_{ij} - \tilde{x}_{ji}) + \mathcal{B} \mathcal{F} d_{i0} a_{i0} (\tilde{x}_{i0} - \tilde{x}_0) + f(x_i, t) - f(x_0, t) \\ d_{ij} &= c_{ij} a_{ij} (\tilde{x}_{ij} - \tilde{x}_{ji})^T \mathcal{M} (\tilde{x}_{ij} - \tilde{x}_{jl}) \\ d_{i0} &= c_{i0} a_{i0} (\tilde{x}_{i0} - \tilde{x}_0)^T \mathcal{M} (\tilde{x}_{i0} - \tilde{x}_0). \end{aligned} \tag{3.30}$$

Theorem 3.2 *Under the condition of Assumptions 3.1 and 3.2, the dynamic edge-based event-triggered consensus control protocol in Eq.* (3.28) *and triggering functions in Eq.* (3.29) *are constructed by Algorithm 3.1, then the leader-follower consensus problem of multiagent system of Eq.* (3.1) *and Eq.* (3.2) *is solved. In addition, the adaptive coupling weights d_{ij} converge to some finite values.*

Proof Consider the following Lyapunov function $\mathcal{V}_2$:

$$\begin{aligned} \mathcal{V}_2 = & \frac{1}{2} \sum_{i \in I[1,N]} \zeta_i^T Q^{-1} \zeta_i + \sum_{i \in I[1,N]} \frac{(d_{i0} - \beta)^2}{4c_{i0}} \\ & + \sum_{i \in I[1,N]} \sum_{j=1, j \neq i}^{N} \frac{(d_{ij} - \beta)^2}{8c_{ij}} \end{aligned} \tag{3.31}$$

where Q is given in Algorithm 3.1 and $\beta = \max\{\frac{1}{\iota}, \frac{4\varrho}{\lambda_1}\}$. Calculate the derivative of the first term of Eq. (3.31) as

$$\begin{aligned} \dot{\mathcal{V}}_{21} = & \sum_{i \in I[1,N]} \zeta_i^T Q^{-1} \Big(\mathcal{A}\zeta_i + \mathcal{BF} \sum_{j \in I[1,N]} d_{ij} a_{ij} (\tilde{x}_{ij} - \tilde{x}_{ji}) \\ & + \mathcal{BF} d_{i0} a_{i0} (\tilde{x}_{i0} - \tilde{x}_0) + f(x_i, t) - f(x_0, t) \Big) \\ = & \sum_{i \in I[1,N]} \zeta_i^T Q^{-1} \mathcal{A} \zeta_i - \sum_{i \in I[1,N]} \sum_{j \in I[1,N]} d_{ij} a_{ij} \zeta_i^T \mathcal{M} (\tilde{x}_{ij} - \tilde{x}_{ji}) \\ & - \sum_{i \in I[1,N]} d_{i0} a_{i0} \zeta_i^T \mathcal{M} (\tilde{x}_{i0} - \tilde{x}_0) \\ & + \sum_{i \in I[1,N]} \zeta_i^T Q^{-1} \big(f(x_i, t) - f(x_0, t) \big). \end{aligned} \tag{3.32}$$

Consider the second term of Eq. (3.32) by measurement errors in Eq. (3.10) and Lemma 1.6 as

$$\begin{aligned} & - \sum_{i \in I[1,N]} \sum_{j \in I[1,N]} d_{ij} a_{ij} \zeta_i^T \mathcal{M} (\tilde{x}_{ij} - \tilde{x}_{ji}) \\ = & - \frac{1}{4} \sum_{i \in I[1,N]} \sum_{j \in I[1,N]} d_{ij} a_{ij} (\tilde{x}_{ij} - \tilde{x}_{ji})^T \mathcal{M} (\tilde{x}_{ij} - \tilde{x}_{ji}) \\ & - \frac{1}{4} \sum_{i \in I[1,N]} \sum_{i \in I[1,N]} d_{ij} a_{ij} (\zeta_i - \zeta_j)^T \mathcal{N} (\zeta_i - \zeta_j) \\ & + \frac{1}{4} \sum_{i \in I[1,N]} \sum_{i \in I[1,N]} d_{ij} a_{ij} (e_{ij} - e_{ji})^T \mathcal{M} (e_{ij} - e_{ji}) \\ \leq & - \frac{1}{4} \sum_{i \in I[1,N]} \sum_{j \in I[1,N]} d_{ij} a_{ij} (\tilde{x}_{ij} - \tilde{x}_{ji})^T \mathcal{M} (\tilde{x}_{ij} - \tilde{x}_{ji}) \\ & + \sum_{i \in I[1,N]} \sum_{j \in I[1,N]} d_{ij} a_{ij} e_{ij}^T \mathcal{M} e_{ij} \\ & - \frac{1}{4} \sum_{i \in I[1,N]} \sum_{j \in I[1,N]} d_{ij} a_{ij} (\zeta_i - \zeta_j)^T \mathcal{M} (\zeta_i - \zeta_j). \end{aligned}$$

Consider the third term of Eq. (3.32) by measurement errors in Eq. (3.10) and Lemma 1.6

$$
\begin{aligned}
&- \sum_{i\in I[1,N]} d_{i0}a_{i0}\zeta_i^T \mathcal{M}(\tilde{x}_{i0}-\tilde{x}_0) \\
&= -\frac{1}{2}\sum_{i\in I[1,N]} d_{i0}a_{i0}\zeta_i^T \mathcal{M}\zeta_i - \frac{1}{2}\sum_{i\in I[1,N]} d_{i0}a_{i0}(\tilde{x}_{i0}-\tilde{x}_0)^T \mathcal{M}(\tilde{x}_{i0}-\tilde{x}_0) \\
&+\frac{1}{2}\sum_{i\in I[1,N]} d_{i0}a_{i0}(e_{i0}-e_0)^T \mathcal{M}(e_{i0}-e_0) \\
&\le -\frac{1}{2}\sum_{i\in I[1,N]} d_{i0}a_{i0}\zeta_i^T \mathcal{M}\zeta_i - \frac{1}{2}\sum_{i\in I[1,N]} d_{i0}a_{i0}(\tilde{x}_{i0}-\tilde{x}_0)^T \mathcal{M}(\tilde{x}_{i0}-\tilde{x}_0) \\
&+\sum_{i\in I[1,N]} d_{i0}a_{i0}e_{i0}^T \mathcal{M} e_{i0} + \sum_{i\in I[1,N]} d_{i0}a_{i0}e_0^T \mathcal{M} e_0.
\end{aligned} \tag{3.33}
$$

Therefore, one can obtain that $\mathcal{V}_{21}$ satisfies

$$
\begin{aligned}
\dot{\mathcal{V}}_{21} \le & \frac{1}{2}\sum_{i\in I[1,N]} \zeta_i^T (Q^{-1}\mathcal{A} + \mathcal{A}Q^{-1} + \alpha^2 Q^{-2} + I_n)\zeta_i \\
& -\frac{1}{4}\sum_{i\in I[1,N]}\sum_{j\in I[1,N]} d_{ij}a_{ij}(\tilde{x}_{ij}-\tilde{x}_{ji})^T \mathcal{M}(\tilde{x}_{ij}-\tilde{x}_{ji}) \\
& +\sum_{i\in I[1,N]}\sum_{j\in I[0,N]} d_{ij}a_{ij}e_{ij}^T \mathcal{M} e_{ij} \\
& -\frac{1}{2}\sum_{i\in I[1,N]} d_{i0}a_{i0}(\tilde{x}_{i0}-\tilde{x}_0)^T \mathcal{M}(\tilde{x}_{i0}-\tilde{x}_0) \\
& +\sum_{i\in I[1,N]} d_{i0}a_{i0}e_0^T \mathcal{M} e_0.
\end{aligned} \tag{3.34}
$$

Calculate the derivative for the third term of Eq. (3.31) as

$$
\dot{\mathcal{V}}_{22} = \sum_{i\in I[1,N]}\sum_{j\in I[1,N]} \frac{d_{ij}-\beta}{4} a_{ij}(\tilde{x}_{ij}-\tilde{x}_{ji})^T \mathcal{M}(\tilde{x}_{ij}-\tilde{x}_{ji}). \tag{3.35}
$$

Then, consider the following formula as

$$
\begin{aligned}
&-\frac{\beta}{8}\sum_{i\in I[1,N]}\sum_{j\in I[1,N]} a_{ij}(\tilde{x}_{ij}-\tilde{x}_{ji})^T\mathcal{M}(\tilde{x}_{ij}-\tilde{x}_{ji})\\
&=-\frac{\beta}{8}\sum_{i\in I[1,N]}\sum_{j\in I[1,N]} a_{ij}(\zeta_i-\zeta_j)^T\mathcal{M}(\zeta_i-\zeta_j)\\
&\quad-\frac{\beta}{8}\sum_{i\in I[1,N]}\sum_{j\in I[1,N]} a_{ij}(e_{ij}-e_{ji})^T\mathcal{M}(e_{ij}-e_{ji})\\
&\quad-\frac{\beta}{4}\sum_{i\in I[1,N]}\sum_{j\in I[1,N]} a_{ij}(e_{ij}-e_{ji})^T\mathcal{M}(\zeta_i-\zeta_j)\\
&\le-\frac{\beta}{8}\sum_{i\in I[1,N]}\sum_{j\in I[1,N]} a_{ij}\zeta_i^T\mathcal{M}(\zeta_i-\zeta_j)\\
&\quad+\frac{\beta}{2}\sum_{i\in I[1,N]}\sum_{j\in I[1,N]} a_{ij}e_{ij}^T\mathcal{M}e_{ij}.
\end{aligned}
$$

It follows from Eq. (3.3.3) that Eq. (3.35) can be written as

$$
\begin{aligned}
\dot{\mathcal{V}}_{22}\le&\left(\frac{1}{4}-\frac{\beta}{8}\right)\sum_{i\in I[1,N]}\sum_{j\in I[1,N]} a_{ij}(\tilde{x}_{ij}-\tilde{x}_{ji})^T\mathcal{M}(\tilde{x}_{ij}-\tilde{x}_{ji})\\
&-\frac{\beta}{8}\sum_{i\in I[1,N]}\sum_{j\in I[1,N]} a_{ij}\zeta_i^T\mathcal{M}(\zeta_i-\zeta_j)\\
&+\frac{\beta}{2}\sum_{i\in I[1,N]}\sum_{j\in I[1,N]} a_{ij}e_{ij}^T\mathcal{M}e_{ij}.
\end{aligned}
\tag{3.36}
$$

Calculate the derivative for the second term of Eq. (3.31) as

$$
\dot{\mathcal{V}}_{23}=\sum_{i\in I[1,N]}\frac{d_{i0}-\beta}{2}a_{i0}(\tilde{x}_{i0}-\tilde{x}_0)^T\mathcal{M}(\tilde{x}_{i0}-\tilde{x}_0).
\tag{3.37}
$$

Then, consider the following formula as

$$
\begin{aligned}
&-\frac{\beta}{4}\sum_{i\in I[1,N]} a_{i0}(\tilde{x}_{i0}-\tilde{x}_0)^T\mathcal{M}(\tilde{x}_{i0}-\tilde{x}_0)\\
&=-\frac{\beta}{4}\sum_{i\in I[1,N]} a_{i0}\zeta_i^T\mathcal{M}\zeta_i-\frac{\beta}{4}\sum_{i\in I[1,N]} a_{i0}(e_{i0}-e_0)^T\mathcal{M}(e_{i0}-e_0)\\
&\quad-\frac{\beta}{2}\sum_{i\subset I[1,N]} a_{i0}\zeta_i^T\mathcal{M}(e_{i0}-e_0)\\
&\le-\frac{\beta}{8}\sum_{i\in I[1,N]} a_{i0}\zeta_i^T\mathcal{M}\zeta_i+\frac{\beta}{2}\sum_{i\in I[1,N]} a_{i0}(e_{i0}^T\mathcal{M}e_{i0}+e_0^T\mathcal{M}e_0).
\end{aligned}
\tag{3.38}
$$

It follows from Eq. (3.38) that Eq. (3.37) satisfies

$$\begin{aligned}\dot{\mathcal{V}}_{23} \leq &-\frac{\beta}{4}\sum_{i\in I[1,N]} a_{i0}(\tilde{x}_{i0}-\tilde{x}_0)^T\mathcal{M}(\tilde{x}_{i0}-\tilde{x}_0)\\ &+\frac{1}{2}\sum_{i\in I[1,N]} d_{i0}a_{i0}(\tilde{x}_{i0}-\tilde{x}_0)^T\mathcal{M}(\tilde{x}_{i0}-\tilde{x}_0)\\ &-\frac{\beta}{8}\sum_{i\in I[1,N]} a_{i0}\zeta_i^T\mathcal{M}\zeta_i+\frac{\beta}{2}\sum_{i\in I[1,N]} a_{i0}(e_{i0}^T\mathcal{M}e_{i0}+e_0^T\mathcal{M}e_0).\end{aligned}$$

Combining Eqs. (3.34), (3.36), and (3.3.3), one obtains that the derivative of $\mathcal{V}_2$ satisfies

$$\begin{aligned}\dot{\mathcal{V}}_2 \leq &\frac{1}{2}\sum_{i\in I[1,N]} \zeta_i^T(Q^{-1}\mathcal{A}+\mathcal{A}Q^{-1}+\alpha^2Q^{-2}+I_n)\zeta_i\\ &-\frac{\beta}{8}\sum_{i\in I[1,N]}\sum_{j\in I[0,N]} a_{ij}\zeta_i^T\mathcal{M}(\zeta_i-\zeta_j)\\ &+\frac{\beta}{2}\sum_{i\in I[1,N]}\sum_{j\in I[1,N]} a_{ij}\{(1+\frac{2}{\beta\iota}\iota d_{ij})e_{ij}^T\mathcal{M}e_{ij}\\ &-\frac{1}{4}(\tilde{x}_{ij}-\tilde{x}_{ji})^T\mathcal{M}(\tilde{x}_{ij}-\tilde{x}_{ji})\}\\ &+\frac{\beta}{2}\sum_{i\in I[1,N]} a_{i0}\big\{(1+\frac{2}{\beta\iota}\iota d_{i0})e_{i0}^T\mathcal{M}e_{i0}-\frac{1}{2}(\tilde{x}_{i0}-\tilde{x}_0)^T\mathcal{M}(\tilde{x}_{i0}-\tilde{x}_0)\big\}\\ &+\frac{\beta}{2}\sum_{i\in I[1,N]} a_{i0}(1+\frac{2}{\beta\iota}\iota d_{i0})e_0^T\mathcal{M}e_0.\end{aligned} \tag{3.39}$$

Simplify Eq. (3.39) into Eq. (3.40) using triggering mechanism composed of f Eq. (3.6) and Eq. (3.29),

$$\begin{aligned}\dot{\mathcal{V}}_2 \leq &\frac{1}{2}\sum_{i\in I[1,N]} \zeta_i^T(Q^{-1}\mathcal{A}+\mathcal{A}Q^{-1}+\alpha^2Q^{-2}+I_n)\zeta_i\\ &-\frac{\beta}{8}\sum_{i\in I[1,N]}\sum_{j\in I[0,N]} a_{ij}\zeta_i^T\mathcal{M}(\zeta_i-\zeta_j)+\bar{\varpi}(t)\end{aligned} \tag{3.40}$$

where $\bar{\varpi}(t)=\frac{\beta}{2}\sum_{i\in I[1,N]}\sum_{j\in I[0,N]} a_{ij}v_{ij}e^{-\omega_{ij}t}+\frac{\beta}{2}\sum_{i\in I[1,N]} a_{i0}(1+2\iota d_{i0})$ $v_0e^{-\omega_0 t}$ is given. Equation (3.40) holds for each interval $[t_k, t_{k+1})$, $\{t_k : k\in\mathbb{N}\}$ is the strictly increasing sequence of unique elements, consisting of elements in $\{t_k^0\cup t_k^{ij} : k\in\mathbb{N}\}$, for all $i\in I[1,N]$, $j\in I[0,N]$.

Similar to Eq. (3.25), one obtains

$$\dot{\mathcal{V}}_2 \leq \frac{1}{2}\hat{\zeta}^T\big(I_N \otimes (\mathcal{A}Q + Q\mathcal{A}^T + \alpha^2 I_n + Q^2) - \frac{\beta}{4}(\mathcal{H} \otimes \mathcal{B}\mathcal{B}^T)\big)\hat{\zeta} + \bar{\varpi}(t) \tag{3.41}$$

and

$$\begin{aligned}\dot{\mathcal{V}}_2 &\leq \frac{1}{2}\sum_{i\in I[1,N]} \tilde{\zeta}_i(\mathcal{A}Q + Q\mathcal{A}^T + \alpha^2 I_n + Q^2 - \frac{\beta}{4}\lambda_1\mathcal{B}\mathcal{B}^T)\tilde{\zeta}_i + \bar{\varpi}(t)\\ &= \frac{1}{2}\sum_{i\in I[1,N]} \tilde{\zeta}_i\bar{\Omega}\tilde{\zeta}_i + \bar{\varpi}(t).\end{aligned} \tag{3.42}$$

In light of the definition of β, Lemma 1.8, and Algorithm 3.1, one can obtain

$$\bar{\Omega} \leq \mathcal{A}Q + Q\mathcal{A}^T + \alpha^2 I_n + Q^2 - \varrho\mathcal{B}\mathcal{B}^T < 0. \tag{3.43}$$

Therefore, there exists a positive definite matrix $\bar{\mathcal{P}} > 0$ such that $\bar{\Omega} \leq -\bar{\mathcal{P}} < 0$ and

$$\dot{\mathcal{V}}_2 \leq -\frac{1}{2}\sum_{i\in I[1,N]} \tilde{\zeta}_i^T\bar{\mathcal{P}}\tilde{\zeta}_i + \bar{\varpi}(t) \leq -\frac{\lambda_{min}(\bar{\mathcal{P}})\lambda_{min}(Q^{-2})}{2}\zeta^T\zeta + \bar{\varpi}(t). \tag{3.44}$$

Subsequently, we analyze that the consensus error ζ eventually approaches zero through the following two cases.

Case 1: If $\frac{\lambda_{min}(\bar{\mathcal{P}})\lambda_{min}(Q^{-2})}{2}\zeta^T\zeta \leq \bar{\varpi}_{ij}(t)$ holds, it is easy to verify that consensus error ζ finally converges to zero since $\lim_{t\to\infty}\bar{\varpi}(t) = 0$. Therefore, it is not difficult to obtain that $\dot{d}_{ij}$ converges to zero, which implies that adaptive coupling weight d_{ij} tends to the finite value.

Case 2: If $\frac{\lambda_{min}(\bar{\mathcal{P}})\lambda_{min}(Q^{-2})}{2}\zeta^T\zeta \geq \bar{\varpi}_{ij}(t)$ holds, it is easy to verify that $\dot{\mathcal{V}}_2 < 0$. Therefore, we can obtain that $\mathcal{V}_2$ is monotonically non-increasing, and $\lim_{t\to\infty}\mathcal{V}_2(t)$ exists. Further, one can obtain that $\zeta,\ d_{ij}$ are bounded based on the construction of $\mathcal{V}_2$. Since d_{ij} is monotonously increasing, one can obtain that the adaptive coupling weight d_{ij} converges to the finite value.

According to Eq. (3.30) and Assumption 3.1, one can obtain that $\dot{\zeta}$ is bounded. It has been proved above that d_{ij} is bounded, so there exists a normal number d such that $d_{ij} \leq \bar{d},\ \forall\, i = 1, \ldots, N,\ j = 0, 1, \ldots, N$.

Consider the following formula:

$$\begin{aligned}\bar{\varpi}(t) &\leq \frac{\beta}{2}\sum_{i\in I[1,N]}\sum_{j\in I[0,N]} a_{ij}\nu_{ij}e^{-\omega_{ij}t} + \frac{\beta}{2}\sum_{i\in I[1,N]} a_{i0}(1 + 2\iota\bar{d})\nu_0 e^{-\omega t}\\ &=: \tilde{\varpi}(t).\end{aligned} \tag{3.45}$$

It follows from Eq. (3.44) and Eq. (3.45) that

$$\int_0^{+\infty} \zeta^T \zeta dt \leq \frac{2}{\lambda_{min}(\bar{\mathcal{P}})\lambda_{min}(Q^{-2})}(\mathcal{V}_2(0) - \mathcal{V}_2(+\infty) + \tilde{\varpi}), \tag{3.46}$$

where $\tilde{\varpi} = \frac{\beta}{2}\sum_{i\in I[1,N]}\sum_{j\in I[0,N]} a_{ij}\frac{v_{ij}}{\omega_{ij}} + \frac{\beta}{2}\sum_{i\in I[1,N]} a_{i0}(1+2\iota\bar{d})\frac{v_0}{\omega_0}$ is given. It follows from the Comparative discrimination method that $\lim_{t\to\infty}\int_0^t \zeta^T(s)\zeta(s)ds$ exists.

Calculate the second derivative of the following formula:

$$\frac{d^2}{dt^2}\int_0^t \zeta^T(s)\zeta(s)ds = 2\zeta^T(t)\dot{\zeta}(t),$$

which implies $\frac{d^2}{dt^2}\int_0^t \zeta^T(s)\zeta(s)ds$ is bounded. According to Lemma 1.5, we can obtain $\lim_{t\to+\infty}\frac{d}{dt}\int_0^t \zeta^T(s)\zeta(s)ds = 0$, which implies that $\lim_{t\to+\infty}\zeta(t) = 0$. Therefore, the leader-follower consensus problem is solved. That is the end of proof.

The following presents a theorem to analyze the feasibility of the proposed triggering mechanism in Eq. (3.6).

Theorem 3.3 *Under the conditions of Theorems 3.1 and 3.2, the multiagent system does not exhibit the Zeno behavior if nonlinear term $f(\cdot)$ is bounded.*

Proof We first analyze that the multiagent system does not exhibit Zeno behavior under Theorem 3.1. The contradiction method is adopted. The proof is divided into two cases.

Firstly, for the leader 0, assume that the Zeno behavior happens at $T_0 < \infty$. Therefore, one obtains that $\lim_{k\to\infty} t_k^0 = T_0$,

$$\forall\, \varepsilon_0 > 0,\ \exists\, N(\varepsilon_0) \text{ such that } T_0 - \varepsilon_0 < t_k^0 \leq T_0,\ \forall\, k \geq N(\varepsilon_0). \tag{3.47}$$

Calculate the derivative of e_0,

$$\dot{e}_0 = \mathcal{A}e_0 - f(x_0, t). \tag{3.48}$$

Then, calculate the right-hand Dini derivative of e_0,

$$D^+\|e_0\| \leq \frac{\|e_0^T\|}{\|e_0\|}\|\dot{e}_0\| = \|\mathcal{A}e_0 - f(x_0, t)\| \leq \|\mathcal{A}\|\|e_0\| + \mu \tag{3.49}$$

where μ is the upper bound of $\|f(x_0, t)\|$.

Combining $e_0(t_k^0) = 0$ and Eq. (3.49), one obtains that

$$\|e_0(t)\| \leq \frac{\mu}{\|\mathcal{A}\|}(e^{\|\mathcal{A}\|(t-t_k^0)} - 1). \tag{3.50}$$

In addition, by using the triggering mechanism in Eq. (3.6) and triggering function in Eq. (3.11), we can know that the condition of the triggering mechanism in Eq. (3.6) does not hold, if the following formula holds,

$$\|e_0\| \le \sqrt{\frac{\upsilon_0}{\|\mathcal{M}\|}e^{-\omega_0 T_0}} \tag{3.51}$$

It follows from Eqs. (3.50) and (3.51) that the adjacent triggering time interval $[t_k^0, t_{k+1}^0)$ satisfies $t_{k+1}^0 - t_k^0 \ge 2\varepsilon_0$ with $\varepsilon_0 = \frac{1}{2\|\mathcal{A}\|}ln(\frac{\|\mathcal{A}\|}{\mu}\sqrt{\frac{\nu_0}{\|\mathcal{M}\|}e^{-\omega_0 T_0}} + 1)$. In light of $t_{k+1}^0 - t_k^0 \ge 2\varepsilon_0$ and (3.47), one obtains that

$$t_{k+1}^0 \ge 2\varepsilon_0 + t_k^0 \ge 2\varepsilon_0 + T_0 - \varepsilon_0 = \varepsilon_0 + T_0,$$

which is contradictory to (3.47). Conclusively, we can conclude that leader will not exhibit Zeno behavior.

For each edge (i, j), $\forall\ i = 0, 1, \ldots, N$, $j = 1, \ldots, N$, assume that Zeno behavior happens at $T_1 < \infty$. Therefore, one has $\lim_{k\to\infty} t_k^{ij} = T_1$, and

$$\forall\, \varepsilon_1 > 0,\ \exists\, N(\varepsilon_1) \text{ such that } T_1 - \varepsilon_1 < t_k^{ij} \le T_1,\ \forall\, k \ge N(\varepsilon_1). \tag{3.52}$$

According to the result of Theorem 3.1, one obtains that $\tilde{x}_i(t_k^{ij}) - \tilde{x}_j(t_k^{ji})$ is bounded for any limited time. Then, calculate the right-hand Dini derivative of $\|e_{ij}(t)\|$.

$$D^+\|e_{ij}(t)\| \le \|\mathcal{A}e_{ij} - d\mathcal{BF}\sum_{j\in I[0,N]} w_{ij}(\tilde{x}_{ij} - \tilde{x}_{ji}) - f(x_i)\| \le \|\mathcal{A}\|\|e_{ij}\| + \Theta_{ij} \tag{3.53}$$

where Θ_{ij} is the upper bound of $\|d\mathcal{BF}\sum_{j\in I[0,N]} a_{ij}(\tilde{x}_{ij} - \tilde{x}_{ji}) + f(x_i)\|$, $\tilde{x}_{0i} = \tilde{x}_0$, $\forall\ i \in I[1, N]$. According to $\|e_{ij}(t_k^{ij})\| = 0$ and Comparison lemma (Khalil 2002), we can simply

$$\|e_{ij}(t)\| \le \frac{\Theta_{ij}}{\|\mathcal{A}\|}(e^{\|\mathcal{A}\|(t-t_k^{ij})} - 1). \tag{3.54}$$

Based on the triggering mechanism in Eq. (3.6) and triggering function in Eq. (3.11), ignoring state-related terms can lead to the conclusion that the triggering condition in Eq. (3.6) will not be satisfied if the following formula holds:

$$\|e_{ij}\| \le \sqrt{\frac{\nu_{ij}}{\|\mathcal{M}\|}e^{-\frac{\omega_{ij}}{2}T_1}}. \tag{3.55}$$

Combining Eqs. (3.54) and (3.55), we conclude that the adjacent triggering time interval $[t_k^{ij}, t_{k+1}^{ij})$ satisfies $t_{k+1}^{ij} - t_k^{ij} \ge 2\varepsilon_1$ with $\varepsilon_1 = \frac{1}{\|\mathcal{A}\|}ln\{\frac{\|A\|}{\Theta_{ij}}(\sqrt{\frac{\nu_{ij}}{\|\mathcal{A}\|}e^{\frac{\omega_{ij}}{2}T_1}}) + 1\}$. According to $t_{k+1}^{ij} - t_k^{ij} \ge 2\varepsilon_1$ and (3.52), one obtains

$$t_{k+1}^{ij} \geq 2\varepsilon_1 + t_k^{ij} \geq 2\varepsilon_1 + T_1 - \varepsilon_1 = \varepsilon_1 + T_1,$$

which is contradictory to (3.52). Conclusively, we can obtain that the leader will not exhibit Zeno behavior. All in all, the Zeno behavior does not exhibit if nonlinear term $f(\cdot)$ is bounded under Theorem 3.1.

Finally, similar to the above proof process, we can analyze that under the condition of Theorem 3.2, the multiagent system will not exhibit Zeno behavior. The detailed proof process has been omitted, the proof is end. □

3.4 Consensus Control for Multiagent Systems with One-Sided Lipschitz Dynamics by the Node-Based Event-Triggered Mechanisms

3.4.1 Leader-Follower Consensus Control by Node-Based Event-Triggered Mechanisms

Construct the following distributed node-based event-triggered control scheme for follower i as

$$\begin{aligned} u_i =& \mathcal{K} \sum_{j\in I[1,N]} d_{ij} w_{ij}(\tilde{x}_i - \tilde{x}_j), \text{ for } i \in I[1, N] \\ \dot{d}_{ij} =& \varepsilon_{ij} w_{ij}(\tilde{x}_i - \tilde{x}_j)^T \mathcal{M}(\tilde{x}_i - \tilde{x}_j), \text{ for } j \in I[1, N], \end{aligned} \tag{3.56}$$

in which d_{ij} ($d_{ij}(0) = d_{ji}(0)$) denotes the time-varying coupling weight, $\varepsilon_{ij} = \varepsilon_{ji} \in R$ denotes the positive constant. $\mathcal{K}$, $\mathcal{M}$ denote the feedback gain matrices to be specified. $\tilde{x}_i(t) = e^{\mathcal{A}(t-t_k^i)} x_i(t_k^i)$. t_k^i is the triggering instant.

Define the state estimation error $e_i(t)$ and consensus error $z_i(t)$ as

$$e_i(t) = \tilde{x}_i(t) - x_i(t), \ t \in [t_k^i, t_{k+1}^i) \tag{3.57}$$

$$z_i(t) = x_i(t) - x_0(t), \ t \in [t_k^i, t_{k+1}^i). \tag{3.58}$$

Design the following node-based event-triggered mechanism:

$$t_{k+1}^i = \inf\{t \geq t_k^i | h_i \geq 0\} \tag{3.59}$$

where h_i is the triggering function to be designed.

$$h_i(t) = \sum_{j\in I[0,N]} (1 + 2\iota d_{ij}) a_{ij} \|\mathcal{M}\| \|e_i\|^2 - \frac{1}{4} \sum_{j\in I[1,N]} a_{ij} (\tilde{x}_i - \tilde{x}_j)^T \mathcal{M}(\tilde{x}_i - \tilde{x}_j)$$

$$- w_{i1}(x_i - x_0)'\mathcal{M}(x_i - x_0) - \psi_i e^{-\omega_i t} \tag{3.60}$$

with $\iota,\ \psi_i, \omega_i \in R$ being the positive constants, and $e_0 = 0$.

Algorithm 3.2 Under Assumptions 3.2–3.4, the related parameters of the distributed controller (3.56) and triggering functions (3.60) are obtained as follows: $\mathcal{K} = -\mathcal{B}^T Q^{-1}$, $\mathcal{M} = Q^{-1}\mathcal{B}\mathcal{B}^T Q^{-1}$, where $Q > 0$ is the solution of (3.61). There exist some scalars $\alpha > 0$, $\beta > 0$ and matrix Q satisfying that

$$\begin{bmatrix} \{\mathcal{A}Q + Q\mathcal{A}^T + (\alpha\varrho + \beta\sigma)Q^2 & `I - \frac{\alpha}{2}Q + \frac{\beta\varsigma}{2}Q \\ -\chi\mathcal{B}\mathcal{B}^T\} & \\ I - \frac{\alpha}{2}Q + \frac{\beta\varsigma}{2}Q & -\beta I \end{bmatrix} < 0 \tag{3.61}$$

where $\chi > 0$, $\varrho, \sigma, \varsigma$ are constants defined in (3.4) and (3.5).

Remark 3.5 According to Lemma 1.8, and cone complementary linearization technique (Majeed et al. 2015), through the simple transformation, we can know that the solution Q of (3.61) can be solved by the following LMIs:

1. If $\alpha\varrho + \beta\sigma = 0$

$$\begin{bmatrix} \mathcal{A}Q + Q\mathcal{A}' - \chi\mathcal{B}\mathcal{B}' & I - \frac{\alpha}{2}Q + \frac{\beta\varsigma}{2}Q \\ I - \frac{\alpha}{2}Q + \frac{\beta\varsigma}{2}Q & -\beta I \end{bmatrix} < 0. \tag{3.62}$$

2. If $\alpha\varrho + \beta\sigma > 0$

$$\begin{bmatrix} \mathcal{A}Q + Q\mathcal{A}' - \chi\mathcal{B}\mathcal{B}' & I - \frac{\alpha}{2}Q + \frac{\beta\varsigma}{2}Q & \sqrt{\alpha\varrho + \beta\sigma}Q \\ I - \frac{\alpha}{2}Q + \frac{\beta\varsigma}{2}Q & -\beta I & 0 \\ \sqrt{\alpha\varrho + \beta\sigma}Q & 0 & -I \end{bmatrix} < 0. \tag{3.63}$$

3. If $\alpha\varrho + \beta\sigma < 0$

$$min\ trace(Q'Q + 0.5\mathcal{P}'\mathcal{P} - 0.5\mathcal{P}'\left(\alpha\varrho + \beta\sigma)Q^2\right)$$
$$xs.t. \left\{ \begin{aligned} &\begin{bmatrix} Q & I \\ I & \mathcal{P}' \end{bmatrix} \geq 0, \begin{bmatrix} \mathcal{P} & I \\ I & \mathcal{P}' \end{bmatrix} \geq 0 \\ &\begin{bmatrix} -(\alpha\varrho + \beta\sigma)I & Q, \\ Q & \mathcal{P}' \end{bmatrix} \geq 0 \end{aligned} \right. \tag{3.64}$$

where $\mathcal{P} = -(\alpha\varrho + \beta\sigma)Q^2$. Apparently, (3.62)–(3.64) are all LMIs, which are easy to solve by MATLAB.

Based on the definition of z_i in (3.58), calculate the derivative of z_i as

$$\begin{aligned} \dot{z}_i &= \mathcal{A}z_i + \mathcal{B}\mathcal{K} \sum_{j \in I[0,N]} d_{ij} w_{ij}(\tilde{x}_i - \tilde{x}_j) + f(x_i) - f(x_0) \\ \dot{d}_{ij} &= \varepsilon_{ij} w_{ij}(\tilde{x}_i - \tilde{x}_j)^T \mathcal{M}(\tilde{x}_i - \tilde{x}_j). \end{aligned} \tag{3.65}$$

The following theorem gives the main result to solve the leader-follower consensus issue.

Theorem 3.4 *Under the condition of Assumptions 3.2–3.4, the node-based event-triggered consensus scheme in Eq.* (3.56) *and triggering function in Eq.* (3.60) *are constructed by Algorithm 3.2, the leader-follower consensus problem of multiagent system in Eq.* (3.1) *is solved, and the adaptive coupling weights* d_{ij} *converge to some finite values. In addition, the multiagent system does not exhibit the Zeno behavior.*

Proof Consider the Lyapunov candidate function as

$$\begin{aligned} V = &\sum_{i\in I[1,N]} z_i^T Q^{-1} z_i + \frac{1}{4} \sum_{i\in I[1,N]} \sum_{j=1, j\neq i}^{N} \frac{(d_{ij}-\tau)^2}{\varepsilon_{ij}} \\ &+ \frac{1}{2} \sum_{i\in I[1,N]} \frac{(d_{i0}-\tau)^2}{\varepsilon_{i1}} \end{aligned} \tag{3.66}$$

where $\tau = max\{\frac{4\chi}{\lambda_{min}(\mathcal{H})}, \frac{1}{\iota}\}$ is given. Calculate the derivative of $\mathcal{V}$ along (3.65) as

$$\begin{aligned} \dot{V} =& 2 \sum_{i\in I[1,N]} z_i^T Q^{-1} \Big(\mathcal{A} z_i + \mathcal{BK} \sum_{j\in I[0,N]} d_{ij} a_{ij} (\tilde{x}_i - \tilde{x}_j) + f(x_i) - f(x_0) \Big) \\ &+ \frac{1}{2} \sum_{i\in I[1,N]} \sum_{j\in I[1,N]} (d_{ij}-\tau) a_{ij} (\tilde{x}_i - \tilde{x}_j)^T \mathcal{M} (\tilde{x}_i - \tilde{x}_j) \\ &+ \sum_{i\in I[1,N]} (d_{i0}-\tau) a_{i1} (\tilde{x}_i - \tilde{x}_0)^T \mathcal{M} (\tilde{x}_i - \tilde{x}_0). \end{aligned} \tag{3.67}$$

According to the definition of e_i and z_i in Eqs. (3.57) and (3.58), we can obtain the following inequality:

$$\begin{aligned} &-2 \sum_{i\in I[1,N]} \sum_{j\in I[1,N]} d_{ij} a_{ij} z_i^T \mathcal{M} (\tilde{x}_i - \tilde{x}_j) \\ \leq& -\frac{1}{2} \sum_{i\in I[1,N]} \sum_{j\in I[1,N]} d_{ij} a_{ij} (\tilde{x}_i - \tilde{x}_j)^T \mathcal{M} (\tilde{x}_i - \tilde{x}_j) \\ &+ 2 \sum_{i\in I[1,N]} \sum_{j\in I[1,N]} d_{ij} a_{ij} e_i^T \mathcal{M} e_i \end{aligned} \tag{3.68}$$

and

$$\begin{aligned} -2 \sum_{i\in I[1,N]} d_{i0} a_{i0} z_i^T \mathcal{M} (\tilde{x}_i - \tilde{x}_0) \leq& - \sum_{i\in I[1,N]} d_{i0} a_{i0} z_i^T \mathcal{M} (\tilde{x}_i - \tilde{x}_0) \\ &+ \sum_{i\in I[1,N]} d_{i0} a_{i0} e_i^T \mathcal{M} e_i. \end{aligned} \tag{3.69}$$

It follows from (3.67) and (3.68) that Eq. (3.69) can be written as

$$\begin{aligned}\dot{V} \leq & 2 \sum_{i\in I[1,N]} z_i^T Q^{-1} z_i - \tau \sum_{i\in I[1,N]} \sum_{j\in I[1,N]} a_{ij} \tilde{x}_i^T \mathcal{M}(\tilde{x}_i - \tilde{x}_j) \\ & - \tau \sum_{i\in I[1,N]} a_{i0} (\tilde{x}_i - \tilde{x}_0)^T \mathcal{M}(\tilde{x}_i - \tilde{x}_0) + 2 \sum_{i\in I[1,N]} \sum_{j\in I[1,N]} d_{ij} a_{ij} e_i^T \mathcal{M} e_i \\ & + \sum_{j\in I[1,N]} d_{i0} a_{i0} e_i^T \mathcal{M} e_i - 2 \sum_{i\in I[1,N]} z_i^T Q^{-1} \Big(f(x_i) - f(x_1) \Big). \end{aligned} \tag{3.70}$$

According to Lemma 1.6, we can obtain that

$$\begin{aligned} & -\frac{\tau}{4} \sum_{i\in I[1,N]} \sum_{j\in I[1,N]} a_{ij} (\tilde{x}_i - \tilde{x}_j)^T \mathcal{M}(\tilde{x}_i - \tilde{x}_j) \\ \leq & -\frac{\tau}{4} \sum_{i\in I[1,N]} \sum_{j\in I[1,N]} a_{ij} z_i^T \mathcal{M}(z_i - z_j) + \tau \sum_{i\in I[1,N]} \sum_{j\in I[1,N]} a_{ij} e_i^T \mathcal{M} e_i. \end{aligned} \tag{3.71}$$

According to the triggering mechanism in (3.59), (3.71), ands (3.70), we can obtain the following inequality:

$$\begin{aligned}\dot{V} \leq & 2 \sum_{i\in I[1,N]} z_i^T Q^{-1} \mathcal{A} z_i - \frac{\tau}{4} \sum_{i\in I[1,N]} \sum_{j\in I[1,N]} a_{ij} z_i^T \mathcal{M}(z_i - z_j) \\ & - 2 \sum_{i\in I[1,N]} z_i^T Q^{-1} \Big(f(x_i) - f(x_0) \Big) \\ & + \tau \sum_{i\in I[1,N]} \{ \sum_{j\in I[1,N]} (1 + \frac{2}{\tau\iota} \iota d_{ij}) a_{ij} e_i^T \mathcal{M} e_i \\ & - \frac{1}{4} \sum_{j\in I[1,N]} a_{ij} (\tilde{x}_i - \tilde{x}_j)^T \mathcal{M}(\tilde{x}_i - \tilde{x}_j) \\ & + \frac{1}{\tau\iota} \iota d_{i0} a_{i0} e_i^T \mathcal{M} e_i - a_{i0} (x_i - x_0)^T \mathcal{M}(x_i - x_0) \} \\ \leq & 2 \sum_{i\in I[1,N]} z_i^T Q^{-1} \mathcal{A} z_i - \frac{\tau}{4} \sum_{i\in I[1,N]} \sum_{j\in I[1,N]} a_{ij} z_i^T \mathcal{M}(z_i - z_j) \\ & - 2 \sum_{i\in I[1,N]} z_i^T Q^{-1} \Big(f(x_i) - f(x_0) \Big) + \tau \sum_{i\in I[1,N]} \psi_i e^{-\omega_i t}. \end{aligned} \tag{3.72}$$

According to Assumptions 3.3 and 3.4, one can obtain that there exist positive constants α, β satisfying the following inequalities:

$$\alpha \varrho \sum_{i\in I[1,N]} z_i^T z_i - \alpha \sum_{i\in I[1,N]} z_i^T \Big(f(x_i) - f(x_0) \Big) \succeq 0, \tag{3.73}$$

$$\beta\sigma \sum_{i\in I[1,N]} z_i^T z_i + \beta\varsigma \sum_{i\in I[1,N]} z_i^T\Big(f(x_i)-f(x_0)\Big) \\ -\beta \sum_{i\in I[1,N]} \|f(x_i)-f(x_0)\|^2 \geq 0. \tag{3.74}$$

Substituting (3.73) and (3.74) into (3.72), one can obtain the following inequality:

$$\begin{aligned}\dot V \leq & 2 \sum_{i\in I[1,N]} z_i'^T Q^{-1}\mathcal{A}z_i - \frac{\tau}{4}\sum_{i\in I[1,N]}\sum_{i\in I[1,N]} d_{ij}a_{ij}z_i^T \mathcal{M}(z_i-z_j)\\ &+(\alpha\varrho+\beta\sigma)\sum_{i\in I[1,N]} z_i^T z_i + \beta\varsigma \sum_{i\in I[1,N]} z_i^T\Big(f(x_i)-f(x_0)\Big)\\ &+2\sum_{i\in I[1,N]} z_i^T Q^{-1}\Big(f(x_i)-f(x_0)\Big) - \alpha \sum_{i\in I[1,N]} z_i^T\Big(f(x_i)-f(x_0)\Big)\\ &-\beta\sum_{i\in I[1,N]} \|f(x_i)-f(x_0)\|^2 + \tau \sum_{i\in I[1,N]} \psi_i e^{-\omega_i t}.\end{aligned} \tag{3.75}$$

Further, we can obtain that (3.75) satisfies that

$$\begin{aligned}\dot V \leq & z^T\Big(I_N\otimes\big(Q^{-1}\mathcal{A}+\mathcal{A}^TQ^{-1}+(\alpha\varrho+\beta\sigma)I\big)\Big)z - \frac{\tau}{4}z^T(\mathcal{H}\otimes\mathcal{M})z\\ &- f^T(\beta I\otimes I)f + z^T\Big(I_N\otimes(2Q^{-1}-\alpha I+\beta\varsigma I)\Big)f\\ &+\tau\sum_{i\in I[1,N]}\psi_i e^{-\omega_i t}\end{aligned} \tag{3.76}$$

where $z = \text{col}(z_1, z_2 \ldots, z_N)$ is given, $f = \text{col}\Big(f(x_1)-f(x_0), \ldots, f(x_N)-f(x_0)\Big)$.

There exists orthogonal matrix U such that $U^T\mathcal{H}U = \text{diag}\{\lambda_1(\mathcal{H}), \ldots, \lambda_N(\mathcal{H})\}$. Then, set $\tilde z = (U\otimes I)z$, $\tilde f = (U\otimes I)f$. Use $\tau \geq \frac{4\chi}{\lambda_{min}(\mathcal{H}_1)}$, and simplify (3.76) as

$$\begin{aligned}\dot V \leq & \tilde z^T\Big(I_N\otimes\big(Q^{-1}\mathcal{A}+\mathcal{A}^TQ^{-1}+(\alpha\sigma+\beta\sigma)I-\chi\mathcal{M}\big)\Big)\tilde z\\ &-\beta\tilde f^T\tilde f + \tilde z^T(I_N\otimes(2Q^{-1}-\alpha I+\beta\varsigma I))\tilde f + \tau\sum_{i\in I[1,N]}\psi_i e^{-\omega_i t}.\end{aligned} \tag{3.77}$$

Let $\tilde z = (\bar U\otimes I)z$, $\tilde f = (\bar U\otimes I)f$, and $\bar U$ is an orthogonal matrix such that $\bar U^T\mathcal{L}\bar U = \text{diag}\{\lambda_1(\mathcal{L}), \ldots, \lambda_N(\mathcal{L})\}$. Since $\xi = \text{col}(\tilde z, \tilde f)$, one can obtain that

$$\dot V \leq \xi^T\Pi\xi + \tau\sum_{i\in I[1,N]}\psi_i e^{-\omega_i t} \tag{3.78}$$

where

$$\Pi = \begin{bmatrix} I_N \otimes \Pi_1 & I_N \otimes \Pi_2 \\ I_N \otimes \Pi_2 & -I_N \otimes \beta I \end{bmatrix} \tag{3.79}$$

and $\Pi_1 = Q^{-1}\mathcal{A} + \mathcal{A}'Q^{-1} + (\alpha\varrho + \beta\sigma)I - \chi\mathcal{M}$, $\Pi_2 = Q^{-1} - \frac{\alpha}{2}I + \frac{\beta\varsigma}{2}I$. After the elementary row and column transformation of (3.79), we can obtain that

$$I_N \otimes \begin{bmatrix} \Pi_1 & \Pi_2 \\ \Pi_2 & -\beta I \end{bmatrix}. \tag{3.80}$$

Obviously, the eigenvalues of matrix (3.79) and matrix (3.80) are the same. It can be obtained by simple calculation that

$$\text{diag}\{Q, I\} \begin{bmatrix} \Pi_1 & \Pi_2 \\ \Pi_2 & -\beta I \end{bmatrix} \text{diag}\{Q, I\} = \Omega \tag{3.81}$$

where

$$\Omega = \begin{bmatrix} \{\mathcal{A}Q + Q\mathcal{A}' + (\alpha\varrho + \beta\sigma)Q^2 & I - \frac{\alpha}{2}Q + \frac{\beta\varsigma}{2}Q \\ -\chi\mathcal{B}\mathcal{B}'\} & \\ I - \frac{\alpha}{2}Q + \frac{\beta\varsigma}{2}Q & -\beta I \end{bmatrix} < 0.$$

Therefore, we can know the matrix Π is negative definite. Let $\mathcal{W} = V + \sum_{i \in I[1,N]} \frac{\tau}{\omega_i}\psi_i e^{-\omega_i t}$. It is clear that $\mathcal{W} \geq 0$. Calculate the derivative of $\mathcal{W}$ as

$$\dot{\mathcal{W}} \leq \xi^T \Pi \xi \leq 0. \tag{3.82}$$

Combining $\mathcal{W} \geq 0$ and $\dot{\mathcal{W}} \leq 0$, one has $\lim_{t\to\infty} \mathcal{W}$ exists and $\mathcal{W}$ is bounded. Therefore, we conclude that the variables d_{ij}, z_i are all bounded. Subsequently, we will analyze that $\lim_{t\to\infty} z_i(t) = 0$. Since $\lim_{t\to\infty} \mathcal{W}$ exists, one can obtain that for $\forall\, \varepsilon_1 > 0$, $\exists\, T_0$, satisfying that $\mathcal{W}(T_1^+) - \mathcal{W}_{12}(T_2^+) < \varepsilon_1, \forall\, T_2 > T_1 > T_0$. Without losing generality, suppose that the triggering instants are ordered over the interval (T_1, T_2), such that $T_1 \leq t_{w1} < t_{w2} < \cdots < t_{wm} \leq T_2$. Based on (3.82), one obtains that

$$\begin{aligned} & -\lambda_{\max}(\Pi) \int_{T_1}^{T_2} \|\xi\|^2 dt \\ \leq & -\int_{T_1}^{T_2} \dot{\mathcal{W}}(t)dt \\ = & -\int_{T_1}^{t_{w1}} \dot{\mathcal{W}}(t)dt - \int_{t_{w1}}^{t_{w2}} \dot{\mathcal{W}}(t)dt - \cdots - \int_{t_{wm}}^{T_2} \dot{\mathcal{W}}(t)dt \\ = & \mathcal{W}(T_1^+) - \mathcal{W}(T_2^+) < \varepsilon_1. \end{aligned} \tag{3.83}$$

It follows from (3.83) that $\lim_{t\to\infty}\lambda_{max}(\Pi)\int_0^t\|\xi\|^2dt$ exists, which implies that $\lim_{t\to\infty}\int_0^t\|\xi\|^2dt$ exists. Conclusively, we obtain that $\lim\limits_{t\to\infty}\int_0^t\|z_i\|^2dt$ exists. Taking advantage of Assumption 3.3, we obtain that $f(x_i)-f(x_1)$ is bounded. It follows from (3.65) that $\dot{z}_i$ is bounded for $t\in[t_w,t_{w+1})$. It follows from $\frac{d}{dt}\|z_i\|^2=\frac{d}{dt}(z_i^Tz_i)=2z_i\dot{z}_i$ that $\frac{d}{dt}\|z_i\|^2$ is bounded. It can be concluded from Lemma 1.5 that $\lim_{t\to\infty}\|z_i\|^2=0$, which implies that $\lim_{t\to\infty}z_i=0$.

Finally, we prove that the multiagent system does not exhibit Zeno behavior under the proposed node-based event-triggered mechanisms. Obviously, there exists $d>0$ such that $d_{ij}\le d$ since d_{ij} is bounded.

Calculate the derivative of $e_i(t)$ as

$$\dot{e}_i(t)=\mathcal{A}e_i(t)-\mathcal{BK}\sum_{j\in I[1,N]}d_{ij}a_{ij}(\tilde{x}_i-\tilde{x}_j)-f(x_i). \tag{3.84}$$

Using $e_i(t_k^i)=0$, we can obtain that

$$\begin{aligned}e_i(t)=&-\mathcal{BK}\int_{t_k^i}^t e^{\mathcal{A}(t-s)}\sum_{j\in I[1,N]}d_{ij}a_{ij}(\tilde{x}_i(s)-\tilde{x}_j(s))ds\\&-\int_{t_k^i}^t e^{\mathcal{A}(t-s)}f(x_i(s))ds.\end{aligned} \tag{3.85}$$

Integrating both sides of (3.85) yields

$$\begin{aligned}\|e_i(t)\|\le&\|\mathcal{BK}\|\|\int_{t_k^i}^t\sum_{j\in I[1,N]}d_{ij}a_{ij}e^{\mathcal{A}(t-s)}(e^{\mathcal{A}(s-t_k^i)}x_i(t_k^i)\\&-e^{\mathcal{A}(s-t_{k'}^j)}x_j(t_{k'}^j))ds\|+\|f(x_i)\|e^{\|\mathcal{A}\|t}\int_{t_k^i}^t e^{-\|\mathcal{A}\|s}ds\\\le&\|\mathcal{BK}\|\|\int_{t_k^i}^t\sum_{j\in I[1,N]}d_{ij}a_{ij}(\tilde{x}_i(t)-\tilde{x}_j(t))ds\|\\&+\|f(x_i)\|e^{\|\mathcal{A}\|(t-t_k^i)}(t-t_k^i)\\\le&\bar{e}_i\triangleq(t-t_k^i)(d\|\mathcal{BK}\|\|\sum_{j\in I[1,N]}a_{ij}(\tilde{x}_i(t)-\tilde{x}_j(t))\|\\&+\|f(x_i)\|e^{\|\mathcal{A}\|(t-t_k^i)}).\end{aligned} \tag{3.86}$$

According to the triggering mechanism (3.59) and the triggering function (3.60), we can obtain that subsystem i will not transmit information to its neighbors until $\bar{e}_i=\tilde{e}_i$ holds, $\tilde{e}_i=\frac{\sqrt{(\sum_{j=1}^N(1+2\iota dw_{ij})\|\mathcal{M}\|\tilde{\varpi}_{ij}}}{(\sum_{j=1}^N(1+2\iota d)w_{ij})\|\mathcal{M}\|}$ and $\tilde{\varpi}_{ij}=\frac{1}{4}\sum_{j=2}^N w_{ij}(\tilde{x}_i-\tilde{x}_j)'\mathcal{M}(\tilde{x}_i-\tilde{x}_j)+w_{i1}(x_i-x_1)'\mathcal{M}(x_i-x_1)+\psi_ie^{-\omega_it}$. Thus, we can obtain that $t-t_k^i$ is positive. Thus, the multiagent system does not exhibit the Zeno behavior under the proposed node-based event-triggered mechanism. The proof is completed. □

3.4.2 Leaderless Consensus Control by Node-Based Event-Triggered Mechanisms

In this subsection, the leaderless consensus problem of nonlinear multiagent system (3.1) is considered by designing node-based event-triggered mechanisms. Construct the adaptive event-triggered control scheme for agent i as

$$\begin{aligned} u_i &= \mathcal{K} \sum_{i \in I[1,N]} d_{ij} a_{ij} (\tilde{x}_i - \tilde{x}_j), \quad i, j \in I[1, N] \\ \dot{d}_{ij} &= \varepsilon_{ij} a_{ij} (\tilde{x}_i - \tilde{x}_j)^T \mathcal{M} (\tilde{x}_i - \tilde{x}_j) \end{aligned} \tag{3.87}$$

where d_{ij} is time-varying coupling weight with $d_{ij}(0) = d_{ji}(0), \varepsilon_{ij} = \varepsilon_{ji} \in R$ is positive constant. The feedback gain matrices $\mathcal{K}, \mathcal{M}$ are defined by Algorithm 3.2. The triggering instant $t_k^i (t_0^i = 0)$ decides on the node-based event-triggered mechanism (3.59), where the triggering function $h_i(t)$ is given by

$$\begin{aligned} h_i(t) = & \sum_{j \in I[1,N]} (1 + 2\iota d_{ij}) a_{ij} \|\mathcal{M}\| \|e_i\|^2 - \psi_i e^{-\omega_i t} \\ & - \frac{1}{4} \sum_{j \in I[1,N]} a_{ij} (\tilde{x}_i - \tilde{x}_j)^T \mathcal{M} (\tilde{x}_i - \tilde{x}_j). \end{aligned} \tag{3.88}$$

Note that $e_i(t)$ is given in (3.57). Construct the consensus error $\eta_i(t)$ as $\eta_i(t) = x_i(t) - \frac{1}{N} \sum_{j \in I[1,N]} x_j(t), \ i \in I[1, N]$.

To achieve the leaderless consensus, the following assumption is given.

Assumption 3.5 The pair $(\mathcal{A}, \mathcal{B})$ is stabilizable, and the topology graph $\mathcal{G}$ is undirected and connected.

In light of (3.1) and (3.87), calculate the derivative of $\eta_i(t)$ as

$$\begin{aligned} \dot{\eta}_i &= \mathcal{A} \eta_i + \mathcal{B} \mathcal{K} \sum_{j \in I[1,N]} d_{ij} a_{ij} (\tilde{x}_i - \tilde{x}_j) + f(x_i) - \frac{1}{N} \sum_{j \in I[1,N]} f(x_j) \\ \dot{d}_{ij} &= \varepsilon_{ij} a_{ij} (\tilde{x}_i - \tilde{x}_j)^T \mathcal{M} (\tilde{x}_i - \tilde{x}_j). \end{aligned} \tag{3.89}$$

The following theorem illustrates the feasibility of the node-based event-triggered consensus protocol designed in this subsection.

Theorem 3.5 *Under the condition of Assumptions 3.3–3.5, the leaderless consensus problem of multiagent system* (3.1) *is solved by the proposed distributed event-triggered control scheme* (3.87) *and the triggering mechanism* (3.56) *with the triggering function* (3.88), *and the time-varying coupling weight d_{ij} converges to finite value. In addition, the multiagent system does not exhibit the Zeno behavior.*

Proof Choose the Lyapunov candidate function as

$$V_1 = \sum_{i\in I[1,N]} \eta_i^T Q^{-1}\eta_i + \frac{1}{4}\sum_{i\in I[1,N]}\sum_{j=1,j\neq i}^{N} \frac{(d_{ij}-\tau_1)^2}{\varepsilon_{ij}} \tag{3.90}$$

where $\tau_1 = \max\{\frac{1}{\iota}, \frac{4\chi}{\lambda_2(\mathcal{L}_1)}\}$ and Q is the solution of (3.61). It is obviously that $V_1 \geq 0$. Then, calculate the derivative of (3.90) subject to (3.89) as

$$\begin{aligned}\dot{V}_1 =& 2\sum_{i\in I[1,N]} \eta_i^T Q^{-1}\mathcal{A}\eta_i - 2\sum_{i\in I[1,N]}\sum_{j\in I[1,N]} a_{ij}\eta_i^T \mathcal{M}(\tilde{x}_i - \tilde{x}_j)\\ &+2\sum_{i\in I[1,N]} \eta_i^T Q^{-1}\Big(f(x_i) - \frac{1}{N}\sum_{j\in I[1,N]} f(x_j)\Big)\\ &+\frac{1}{2}\sum_{i\in I[1,N]}\sum_{j\in I[1,N]} (d_{ij}-\tau_1)a_{ij}(\tilde{x}_i - \tilde{x}_j)^T\mathcal{M}(\tilde{x}_i - \tilde{x}_j).\end{aligned} \tag{3.91}$$

It is easy to verify that $\eta_i = \tilde{x}_i - e_i - \frac{1}{N}\sum_{j\in I[1,N]} x_j$. Thus,

$$\begin{aligned}&-2\sum_{i\in I[1,N]}\sum_{j\in I[1,N]} d_{ij}a_{ij}\eta_i^T\mathcal{M}(\tilde{x}_i - \tilde{x}_j)\\ &\leq -\frac{1}{2}\sum_{i\in I[1,N]}\sum_{j\in I[1,N]} d_{ij}a_{ij}(\tilde{x}_i - \tilde{x}_j)^T\mathcal{M}(\tilde{x}_i - \tilde{x}_j)\\ &\quad+2\sum_{i\in I[1,N]}\sum_{j\in I[1,N]} d_{ij}a_{ij}e_i^T\mathcal{M}e_i.\end{aligned} \tag{3.92}$$

It follows from (3.92) and (3.91) that

$$\begin{aligned}\dot{V}_1 =& 2\sum_{i\in I[1,N]} \eta_i^T Q^{-1}\mathcal{A}\eta_i - \frac{\tau_1}{2}\sum_{i\in I[1,N]}\sum_{i\in I[1,N]} a_{ij}(\tilde{x}_i - \tilde{x}_j)^T\mathcal{M}(\tilde{x}_i - \tilde{x}_j)\\ &+2\sum_{i\in I[1,N]} \eta_i^T Q^{-1}\Big(f(x_i) - \frac{1}{N}\sum_{j\in I[1,N]} f(x_j)\Big).\end{aligned} \tag{3.93}$$

According to Lemma 1.6, one obtains

$$-\frac{\tau_1}{4}\sum_{i\in I[1,N]}\sum_{j\in I[1,N]} a_{ij}(\tilde{x}_i - \tilde{x}_j)^T\mathcal{M}(\tilde{x}_i - \tilde{x}_j)$$

$$\begin{aligned} &\le -\frac{\tau_1}{4} \sum_{i\in I[1,N]} \sum_{j\in I[1,N]} a_{ij}\eta_i^T \mathcal{M}(\eta_i - \eta_j) \\ &\quad + \tau_1 \sum_{i\in I[1,N]} \sum_{j\in I[1,N]} a_{ij} e_i^T M e_i. \end{aligned} \tag{3.94}$$

It follows from (3.93) and (3.94) that

$$\begin{aligned} \dot{V}_1 =& 2 \sum_{i\in I[1,N]} \eta_i^T Q^{-1} \mathcal{A} \eta_i - \frac{\tau_1}{4} \sum_{i\in I[1,N]} \sum_{i\in I[1,N]} a_{ij}\eta_i^T \mathcal{M}(\eta_i - \eta_j) \\ & - \frac{\tau_1}{4} \sum_{i\in I[1,N]} \sum_{j\in I[1,N]} a_{ij}(\tilde{x}_i - \tilde{x}_j)^T \mathcal{M}(\tilde{x}_i - \tilde{x}_j) \\ & + \tau_1 \sum_{i\in I[1,N]} \sum_{j\in I[1,N]} a_{ij} e_i^T M e_i \\ & + 2 \sum_{i\in I[1,N]} \sum_{j\in I[1,N]} d_{ij} a_{ij} e_i^T M e_i \\ & + 2 \sum_{i\in I[1,N]} \eta_i^T Q^{-1} \Big(f(x_i) - \frac{1}{N} \sum_{j\in I[1,N]} f(x_j) \Big). \end{aligned} \tag{3.95}$$

Based on the triggering mechanism (3.56) and triggering function (3.88), one obtains that

$$\begin{aligned} \dot{V}_1 =& 2 \sum_{i\in I[1,N]} \eta_i^T Q^{-1} \mathcal{A} \eta_i - \frac{\tau_1}{4} \sum_{i\in I[1,N]} \sum_{j\in I[1,N]} a_{ij}\eta_i^T \mathcal{M}(\eta_i - \eta_j) \\ & + 2 \sum_{i\in I[1,N]} \eta_i^T Q^{-1} \Big(f(x_i) - \frac{1}{N} \sum_{j\in I[1,N]} f(x_j) \Big) + \tau_1 \sum_{i\in I[1,N]} \psi_i e^{-\omega_i t}. \end{aligned} \tag{3.96}$$

Further, according to (3.73) and (3.74), (3.96) can be written as

$$\begin{aligned} \dot{V}_1 \le& 2 \sum_{i\in I[1,N]} \eta_i^T Q^{-1} \mathcal{A} \eta_i - \frac{\tau_1}{4} \sum_{i\in I[1,N]} \sum_{j\in I[1,N]} a_{ij}\eta_i^T \mathcal{M}(\eta_i - \eta_j) \\ & + (\alpha\varrho + \beta\sigma) \sum_{i\in I[1,N]} \eta_i^T \eta_i + \beta\varsigma \sum_{i\in I[1,N]} \eta_i^T \Big(f(x_i) - \frac{1}{N} \sum_{j\in I[1,N]} f(x_j) \Big) \\ & + 2 \sum_{i\in I[1,N]} \eta_i^T Q^{-1} \Big(f(x_i) - \frac{1}{N} \sum_{j\in I[1,N]} f(x_j) \Big) \\ & - \alpha \sum_{i\in I[1,N]} \eta_i^T \Big(f(x_i) - \frac{1}{N} \sum_{j\in I[1,N]} f(x_j) \Big) \\ & - \beta \sum_{i\in I[1,N]} \| f(x_i) - \frac{1}{N} \sum_{j\in I[1,N]} f(x_j) \|^2 + \tau_1 \sum_{i\in I[1,N]} \psi_i e^{-\omega_i t}. \end{aligned} \tag{3.97}$$

Moreover, (3.97) can be written as

$$\begin{aligned}\dot{V}_1 \leq& \eta^T\Big(I_N \otimes \big((Q^{-1}\mathcal{A} + \mathcal{A}^T Q^{-1}) + (\alpha\varrho + \beta\sigma)I\big) - \frac{\tau_1}{4}(\mathcal{L}\otimes \mathcal{M})\Big)\eta \\ &- F^T(\beta I \otimes I)F + \eta^T(I_N \otimes (2Q^{-1} - \alpha I + \beta\varsigma I)F \\ &+ \tau_1 \sum_{i\in I[1,N]} \psi_i e^{-\omega_i t},\end{aligned} \tag{3.98}$$

where $\eta = \mathrm{col}(\eta_1, \ldots, \eta_N)$, $F = \mathrm{col}\Big(f(x_1) - \frac{1}{N}\sum_{j\in I[1,N]} f(x_j), \ldots, f(x_N) - \frac{1}{N}\sum_{j\in I[1,N]} f(x_j)\Big)$ are given.

Let $\tilde{\eta} = (T \otimes I)\eta$, $\tilde{F} = (T \otimes I)F$, and matrix T satisfies that $T^T\mathcal{L}T = \mathrm{diag}\{0, \lambda_2(\mathcal{L}), \ldots, \lambda_N(\mathcal{L})\}$. Let $\upsilon = col(\tilde{\eta}, \tilde{F})$, and submit it into (3.99). Then, it is easy to obtain that (3.98) satisfies that

$$\begin{aligned}\dot{V}_1 \leq& \tilde{\eta}^T\Big(I_N \otimes \big(Q^{-1}\mathcal{A} + \mathcal{A}^T Q^{-1} + (\alpha\varrho + \beta\sigma)I - \frac{\tau_1}{4}\lambda_2(\mathcal{L})\mathcal{M}\big)\Big)\tilde{\eta} \\ &- \tilde{F}^T(\beta I \otimes I)\tilde{F} + \tilde{\eta}^T\Big(I_N \otimes (2Q^{-1} - \alpha I + \beta\varsigma I)\Big)\tilde{F} + \tau_1 \sum_{i\in I[1,N]} \psi_i e^{-\omega_i t} \\ \leq& \upsilon^T\bar{\Pi}\upsilon + \tau_1 \sum_{i\in I[1,N]} \psi_i e^{-\omega_i t}\end{aligned} \tag{3.99}$$

where $\bar{\Pi}$ is given by $\bar{\Pi} = \begin{bmatrix} I_N \otimes \Pi_1 & I_N \otimes \Pi_2 \\ I_N \otimes \Pi_2 & -I_N \otimes \beta I \end{bmatrix}$. Similar to Theorem 3.3, one can obtain that $\bar{\Pi}$ is the negative definite matrix. The rest is similar to Theorem 3.3, we omit it due to the limited space. The proof is completed. □

3.5 Simulation Examples

Example 3.1 Consider a multiagent system consisting of six followers and one leader, where the followers are labeled as $1, \ldots, 6$, the leader is labeled as 0, and the communication topology is shown in Fig. 3.1. The system dynamic of multiagent system composed of Eqs. (3.1) and (3.2) is given as

$$\mathcal{A} = \begin{bmatrix} 0 & 1 & 0 & 0 \\ -48.6 & -1.25 & 48.6 & 0 \\ 0 & 0 & 0 & 10 \\ 1.95 & 0 & -1.95 & 0 \end{bmatrix},$$

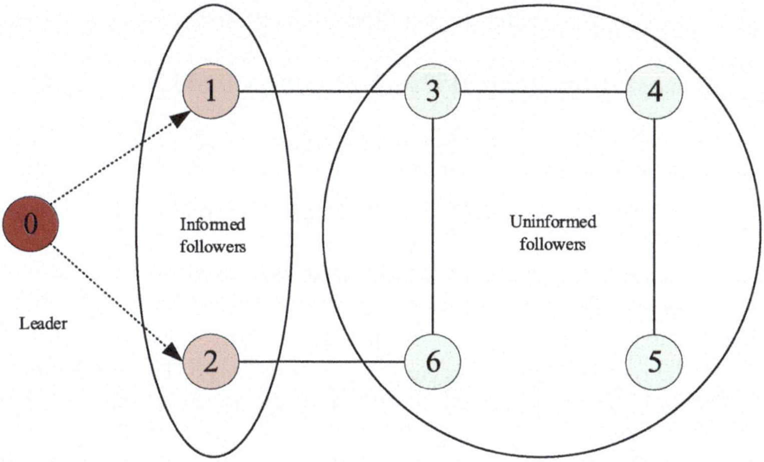

Fig. 3.1 Communication topology $\bar{\mathcal{G}}$

$$\mathcal{B} = \begin{bmatrix} 0 \\ 21.6 \\ 0 \\ 0 \end{bmatrix}, \; f(x_i, t) = \begin{bmatrix} 0 \\ 0 \\ 0 \\ -0.333\sin(x_{i3}) \end{bmatrix}.$$

Obviously, Assumptions 3.1 and 3.2 hold, the Lipschitz constant of Eq. (3.3) in Assumption 3.1 is selected as $\alpha = 0.333$. Select $\varrho = 2.54$, solve the solution of LMI (3.14)

$$Q = \begin{bmatrix} 0.5048 & -0.8909 & 0.4580 & -0.0873 \\ -0.8909 & 23.2183 & -0.0616 & -0.0014 \\ 0.4580 & -0.0616 & 0.5905 & -0.0533 \\ -0.0873 & -0.0014 & -0.0533 & 0.0360 \end{bmatrix}.$$

Then, according to Algorithm 3.1, select the matrices $\mathcal{F}$, $\mathcal{M}$ as

$$\mathcal{F} = \begin{bmatrix} -15.7413 & -1.5094 & 9.9241 & -23.5514 \end{bmatrix},$$

$$\mathcal{M} = \begin{bmatrix} 247.7873 & 23.7598 & -156.2175 & 370.7293 \\ 23.7598 & 2.2783 & -14.9794 & 35.5484 \\ -156.2175 & -14.9794 & 98.4873 & -233.7263 \\ 370.7293 & 35.5484 & -233.7263 & 554.6703 \end{bmatrix}.$$

Randomly select initial states, without losing generality, the initial state values are selected as

$$x_0(0) = \text{col}(0.0975, 0.2785, 0.5469, 0.9575),$$

$$x_1(0) = \text{col}(0.9649, 0.1576, 0.9706, 0.9572),$$

$$x_2(0) = \mathrm{col}(0.4854, 0.8003, 0.1419, 0.4218),$$

$$x_3(0) = \mathrm{col}(0.9157, 0.7922, 0.9595, 0.6557),$$

$$x_4(0) = \mathrm{col}(0.0357, 0.8491, 0.9340, 0.6787),$$

$$x_5(0) = \mathrm{col}(0.7577, 0.7431, 0.3922, 0.6555),$$

$$x_6(0) = \mathrm{col}(0.1712, 0.7060, 0.0318, 0.2769).$$

The other relative parameters are selected as $\iota = 0.1$, $\nu_0 = 3$, $\nu_{ij} = 3$, $\omega_0 = 0.2$, $\omega_{ij} = 0.2$, $c_{ij} = 0.2$, for $i = 1, \ldots, 6$, $j = 0, 1, \ldots, 6$. The simulation results obtained by applying the proposed controller to the given system are shown in Figs. 3.2, 3.3, 3.4 and 3.5.

Figure 3.2 depicts the dynamic response of the adaptive coupling weight, from which we can know the adaptive coupling weight converges to the finite value under Theorem 3.2. Figure 3.3 shows the dynamic response of the consensus error. Obviously, the consensus error eventually tends to stabilize, with the stationary value of 0. Figure 3.4 depicts the dynamic response of the state, from which we can observe that all states eventually tend to stabilize, and not all of their stationary states are zero. Therefore, combining Figs. 3.3 and 3.4, we can know that the proposed control strategy in Eq. (3.28) can achieve consensus for nonlinear multiagent system. Figure 3.5 depicts the triggering instants for the leader and all edges. Obviously, the

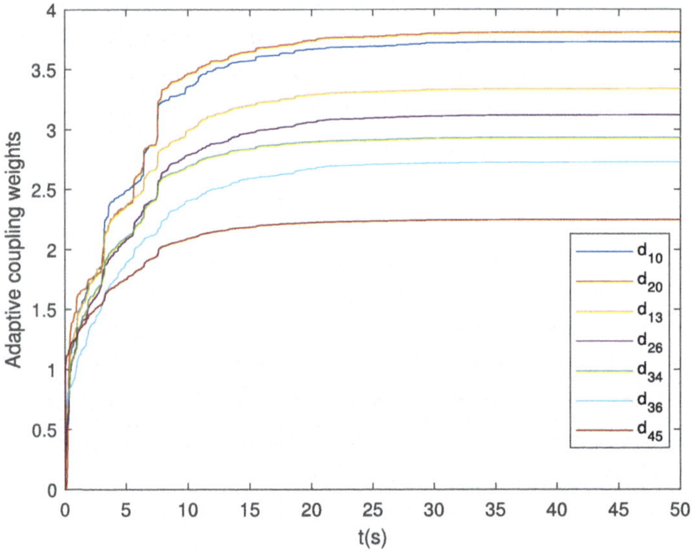

Fig. 3.2 The trajectory of the adaptive coupling weight, reprinted from ref. Zhang et al. (2020b), copyright 2024, with permission from IEEE

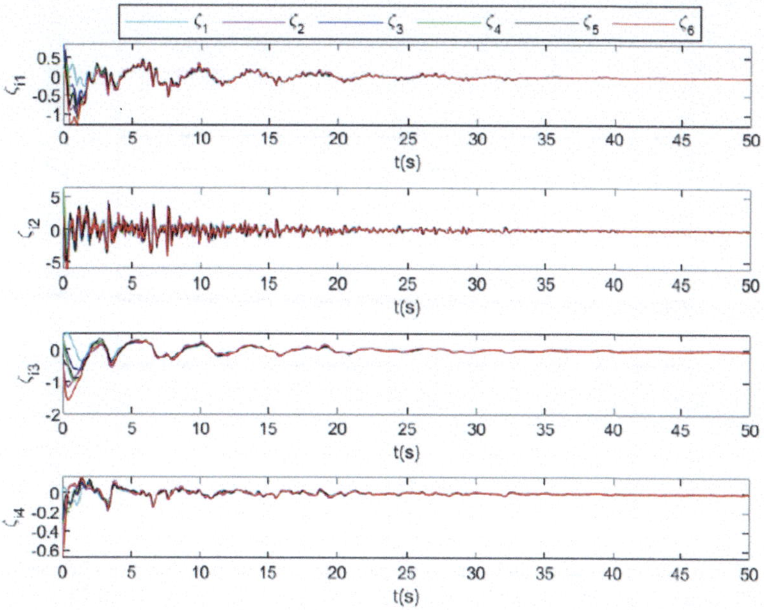

Fig. 3.3 The trajectory of the consensus error, reprinted from ref. Zhang et al. (2020b), copyright 2024, with permission from IEEE

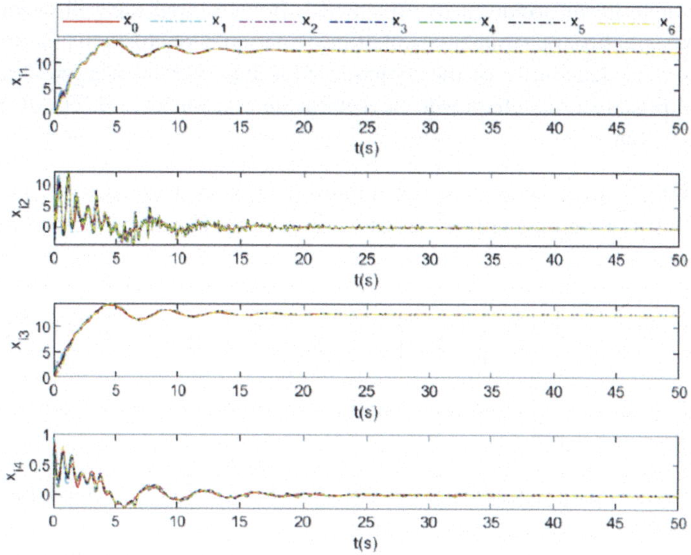

Fig. 3.4 The dynamic response of the state, reprinted from ref. Zhang et al. (2020b), copyright 2024, with permission from IEEE

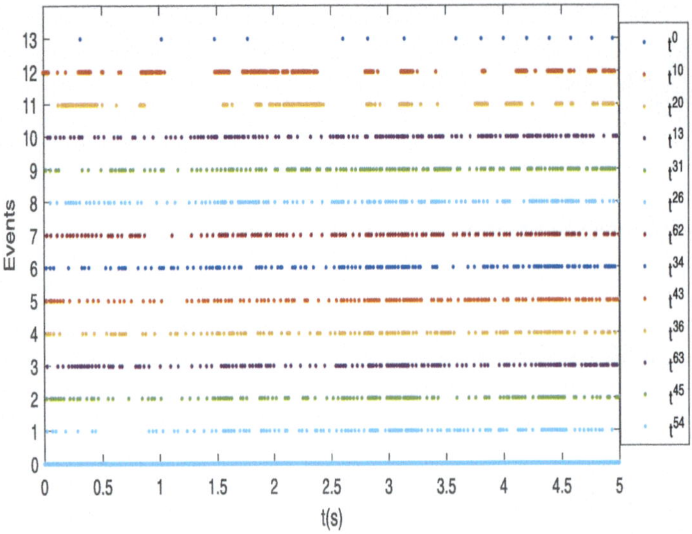

Fig. 3.5 The triggering instants under triggering mechanism composed of Eq. (3.6) and (3.29), reprinted from ref. Zhang et al. (2020b), copyright 2024, with permission from IEEE

multiagent system does not exhibit the Zeno behavior. Moreover, the leader intermittently transmits information to informed followers, and the followers achieve intermittent asynchronous transmission of information. Conclusively, the simulation results verify the feasibility of the dynamic edge-based event-triggered consensus protocol proposed in this chapter to deal with the consensus problem of nonlinear multiagent system.

Example 3.2 This example considers a multiagent system consisting of six agents with the communication method shown in Fig. 3.6. The dynamics of agent i is described as

$$\begin{bmatrix} \dot{x}_{i1} \\ \dot{x}_{i2} \end{bmatrix} = \begin{bmatrix} 1 & -1 \\ 1 & 1 \end{bmatrix} \begin{bmatrix} x_{i1} \\ x_{i2} \end{bmatrix} + \begin{bmatrix} 2 \\ 1 \end{bmatrix} u_i + \begin{bmatrix} x_{i1}(x_{i1}^2 + x_{i2}^2) \\ x_{i2}(x_{i1}^2 + x_{i2}^2) \end{bmatrix}, \ i \in I[1, N].$$

Clearly, the nonlinear function $f(\cdot)$ satisfies Assumptions 3.3 and 3.4, where $\varrho = 0, \sigma = -99, \varsigma = -100$.

Select relative parameters as $\psi_i = 20, \omega_i = 0.5$. Based on Algorithm 3.2, we select the feedback gain matrices $\mathcal{K}, \mathcal{M}$ as

$$\mathcal{K} = -\begin{bmatrix} 2.933 & 1.3535 \end{bmatrix},$$

$$\mathcal{M} = \begin{bmatrix} 8.6044 & 4.5045 \\ 4.5045 & 2.3581 \end{bmatrix}.$$

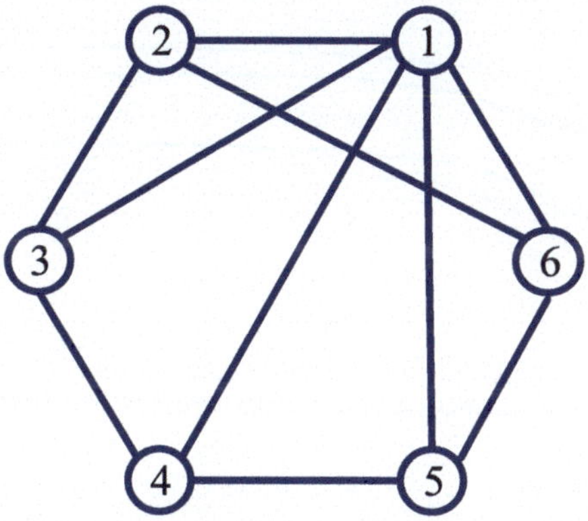

Fig. 3.6 Communication topology, reprinted from ref. Zhang et al. (2021), copyright 2024, with permission from Elsevier

The initial states of multiagent systems are randomly selected and the simulation results are obtained by executing the node-based event-triggered consensus protocol (3.87) which are shown in Figs. 3.7, 3.8, and 3.9.

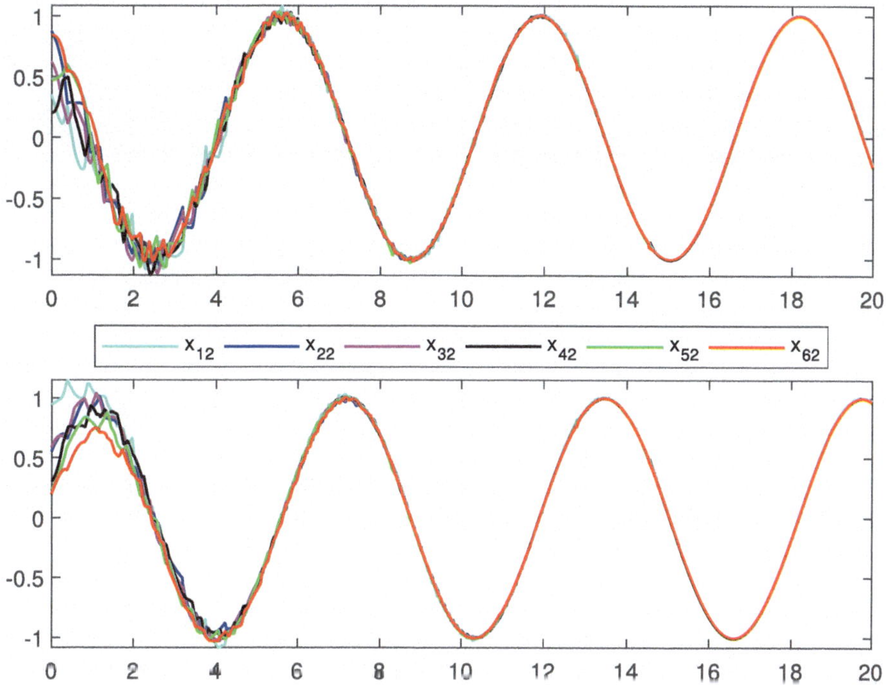

Fig. 3.7 The state trajectory of each agent by Theorem 3.4, reprinted from ref. Zhang et al. (2021), copyright 2024, with permission from Elsevier

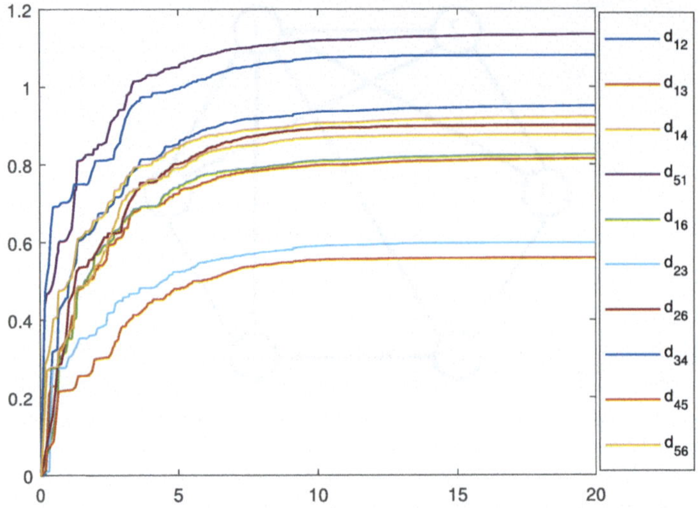

Fig. 3.8 The time-varying coupling weight for each agent by Theorem 3.4, reprinted from ref. Zhang et al. (2021), copyright 2024, with permission from Elsevier

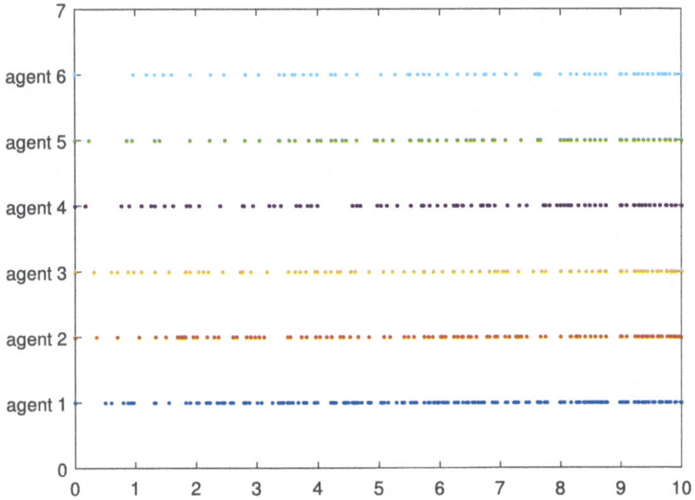

Fig. 3.9 The event instants for each agent by Theorem 3.4, reprinted from ref. Zhang et al. (2021), copyright 2024, with permission from Elsevier

Figure 3.7 depicts the dynamic trajectory of each agent under the proposed node-based event-triggered control protocol (12), and it is clear that the trajectories of the agents eventually converge over time, and thus the leaderless consensus problem is resolved under the conditions of Theorem 3.4. Figure 3.8 depicts the dynamic trajectory of the adaptive weights, from which it can be observed that each adaptive weight eventually converges to a finite value. Figure 3.9 portrays the triggering moments of each intelligence under the proposed node-based event-triggering mechanism, from which it can be observed that Zeno behavior does not occur in multiagent systems. Thus Example 3.2 validates the effectiveness of the proposed node-based event-triggered consensus protocol.

Figure 3.7 depicts the dynamic trajectory of each agent under the proposed node-based event-triggered control protocol (3.87), and it is clear that the trajectories of the agents eventually converge to a common state over time, and thus the leaderless consensus problem is resolved under the conditions of Theorem 3.4. Figure 3.8 depicts the dynamic trajectory of the adaptive weights, from which it can be observed that each adaptive weight eventually converges to the finite value. Figure 3.9 portrays the triggering instant of each agent under the proposed node-based event-triggered mechanism, from which it can be observed that Zeno behavior does not occur in multiagent systems. Thus, Example 3.2 validates the effectiveness of the proposed node-based event-triggered consensus protocol.

3.6 Summary

This chapter first investigated the leader-follower consensus problem for a class of Lipschitz nonlinear multiagent system. Compared with existing relative results, the method proposed in this chapter effectively avoids the continuous transmission of information by all agents, including the leaders, and effectively alleviates the pressure caused by limited communication bandwidth. We have designed node-based event-triggered mechanism and edge-based event-triggered mechanism, respectively. Based on the designed triggering mechanisms, we have designed consensus protocols and combined Schur's complement lemma to provide the conditions for the existence of the controller and triggering mechanism in the form of LMI. The proposed method was applied to the multiagent system under consideration, solving the problem of leader-follower consensus. The introduction of adaptive dynamic event-triggered mechanism ensures that the design of the control protocol and corresponding triggering mechanism does not rely on the global information of the communication topology. Secondly, the leader-follower consensus problem and the leaderless consensus problem were investigated for a class of one-sided Lipschitz nonlinear multiagent systems, where the node-based event-triggered mechanism is designed. In contrast to the edge-based event-triggered mechanism, the node-based event-triggered mechanism needs to design a small number of triggering mechanisms, which is equal to the number of nodes. The number of triggering mechanisms designed by the edge-based event-triggered mechanism is equal to the number of edges. Obviously, the number of the edges is greater than the number of the nodes; for this reason, the edge-based event-triggered mechanism requires high system hardware conditions and should be weighed in the actual use of the choice of which type of triggering mechanism to design. Finally, two examples were introduced to verify the effectiveness of the designed edge-based event-triggered consensus protocol and node-based even-triggered consensus protocol.

References

Ahmad S, Rehan M (2016) On observer-based control of one-sided Lipschitz systems. J Franklin Inst 353:903–916

Cao M, Xiao F, Wang L (2016) Second-order consensus in time-delayed networks based on periodic edge-event driven control. Syst Control Lett 96:37–44

Cheng B, Li Z (2019a) Fully distributed event-triggered protocols for linear multiagent networks. IEEE Trans Autom Control 64(4):1655–1662

Cheng B, Li Z (2019b) Coordinated tracking control with asynchronous edge-based event-triggered communications. IEEE Trans Autom Control 64(10):4321–4328

Cheng B, Wu Z, Li Z (2021) Distributed edge-based event-triggered formation control. IEEE Trans Cybern 51(3):1241–1252

Chu H, Liu X, Zhang W, Cai Y (2016) Observer-based consensus tracking of multi-agent systems with one-sided Lipschitz nonlinearity. J Franklin Inst 353(7):1594–1614

Khalil HK (2002) Nonlinear systems, 3rd edn. Prentice-Hall, Englewood Cliffs, NJ, USA

Li Z, Ren W, Liu X, Fu M (2013) Consensus of multi-agent systems with general linear and Lipschitz nonlinear dynamics using distributed adaptive protocols. IEEE Trans Autom Control 58(7):1786–1791

Majeed R, Ahmad S, Rehan M (2015) Delay-range-dependent observer-based control of nonlinear systems under input and output time-delays. Appl Math Comput 262:145–159

Meng X, Xie L, Soh YC (2018) Event-triggered output regulation of heterogeneous multiagent networks. IEEE Trans Autom Control 63(12):4429–4434

Rehan R, Ahn C, Mustafa G, Ahmad S (2019) Adaptive distributed consensus control of one-sided Lipschitz nonlinear multiagents. IEEE Trans Syst Man Cybern Syst 49(3):568–577

Xiao F, Meng X, Chen T (2015) Sampled-data consensus in switching networks of integrators based on edge events. Int J Control 88(2):391–402

Yang R, Zhang H, Feng G, Yan H, Wang Z (2019) Robust cooperative output regulation of multi-agent systems via adaptive event-triggered control. Automatica 102:129–136

Zhang J, Dai J, Zhang H, Sun S (2020a) Cooperative output regulation of heterogeneous linear multi-agent systems based on the event-triggered distributed control under switching topologies. Appl Math Comput 390:125611

Zhang H, Zhang J, Cai Y, Sun S, Sun J (2020b) Leader-following consensus for a class of nonlinear multiagent systems under event-triggered and edge-event triggered mechanisms. IEEE Trans Cybern 1–12

Zhang J, Zhang H, Cai Y, Wang W (2021) Consensus control for nonlinear multi-agent systems with event-triggered communications. Appl Math Comput 408(6):126341

Chapter 4
Output Regulation of Multiagent Systems Under Non-ideal Environments

4.1 Introduction

It should be noted that the previous chapters are based on the assumption that communication networks are secure, which does not take into account the possibility of malicious attacks during information transmission, leading to information loss. In fact, the process of information transmission is likely to be attacked by malicious attackers, also known as cyber attacks. Cyber attacks launched by attackers usually include two cases: deception attacks (He et al. 2020) and DoS attacks (Zhao et al. 2021). Deception attacks mainly tamper with data in communication networks, while DoS attacks mainly prevent information transmission in communication networks. So far, there have been many research works on DoS attacks (Chen et al. 2019; Teixeira et al. 2015; Feng and Hu 2020; Yang et al. 2019; Xu et al. 2020). For example, the authors in Chen et al. (2019) study security control against DoS attacks in a centralized manner. The authors in Yang et al. (2021) investigate security control against DoS attacks in a distributed manner. To our knowledge, there is limited research on distributed adaptive security issues in multiagent systems with DoS attacks. For multiagent systems with large-scale networks, designing the distributed adaptive security control protocols is crucial.

In addition, the previous chapters assumed that there is no fault in the actuator, without considering actuator faults. DoS attacks are usually caused by adversaries at the network layer, but there can also be actuator faults, which can be caused by abnormal operation of the physical layer or aging materials. Actuator failures can lead to system destabilization. Up to now, there have been some fault-tolerant control protocols that have been proposed in Jiang et al. (2022), Ye et al. (2019), Deng et al. (2021), and Liu et al. (2022) to make up for the influence of actuator faults on system stability. For example, the distributed adaptive event-triggered fault-tolerant control protocol is designed to solve the consensus problem of homogeneous linear multiagent systems in Ye et al. (2019). The novel fault-tolerant controller is proposed in Deng et al. (2021) to solve the output regulation problem of heterogeneous linear multiagent

J. Zhang et al., *Adaptive Collaborative Control of Multiagent Systems by Event-Triggered Mechanisms*, Intelligent Control and Learning Systems 14,
https://doi.org/10.1007/978-981-97-7866-9_4

systems, where actuator faults and DoS attacks coexist. Moreover, the authors in Jiang et al. (2022) design the dual adaptive dynamic event-triggered mechanism, which ensures intermittent communication between neighbors and intermittent updates of controllers and solves the bipartite output consensus in heterogeneous linear multiagent systems with actuator faults. Although event-triggered control can ensure intermittent communication between neighbors, it cannot guarantee asynchronous communication. The authors in Zhang et al. (2021, 2023a), Li et al. (2022), and Cheng and Li (2019) propose the edge-based event-triggered control schemes, when the triggering mechanism is satisfied, the two agent nodes composing of the edge carry out one-way information transmission. The above discussion motivates us to conduct the research about the output regulation problem under DoS attacks and actuator faults.

4.2 Problem Formulation

4.2.1 System Model

Consider the heterogeneous linear multiagent system composed of N followers and one leader. The dynamic model of the ith follower and the leader under consideration in this chapter is

$$\begin{cases} \dot{x}_i = A_i x_i + B_i u_i^f + E_i x_0 \\ y_i = C_i x_i \end{cases} \tag{4.1}$$

and

$$\begin{cases} \dot{x}_0 = A_0 x_0 \\ y_0 = C_0 x_0 \end{cases} \tag{4.2}$$

where $i \in I[1, N]$, $x_i \in R^{n_i}$, $u_i^f \in R^{m_i}$, $y_i \in R^s$ represent state, real input, and output for follower i, $x_0 \in R^{n_0}$, $y_0 \in R^s$ represent state and output for the leader 0, $A_i \in R^{n_i \times n_i}$, $B_i \in R^{n_i \times m_i}$, $E_i \in R^{n_i \times n_0}$, $C_i \in R^{s \times n_i}$ are follower system matrices, and $A_0 \in R^{n_0 \times n_0}$, $C_0 \in R^{s \times n_0}$ are leader system matrices with all eigenvalues of A_0 being semisimple with zero real parts. Note that in this chapter, the leader provides not only perturbation signals but also tracking signals.

The objectives considered in this chapter are as follows.

Output regulation problem: Consider the multiagent system composed of (4.1) and (4.2) under DoS attacks, actuator faults, and communication topology $\bar{\mathcal{G}}$, design the resilient distributed event-triggered fault-tolerant controller to meet the following conditions:

1. When $x_0 = 0$, the closed-loop system consisting of the controller is ultimately bounded;

2. For any $\varepsilon > 0$, there exists $\tau_1 > 0$ such that the regulation output error meets the condition, i.e., $\lim\ \sup_{t\to\infty} \|y_i - y_0\| \le \tau_1\varepsilon$.

To solve the output regulation problem of the multiagent system consisting of (4.1) and (4.2), we give the following assumptions.

Assumption 4.1 The communication topology $\bar{\mathcal{G}}$ includes a directed spanning tree and the leader is its root. The subgraph $\mathcal{G}$ among the followers is undirected.

Assumption 4.2 The pairs (A_i, B_i, C_i) are stabilizable and detectable, for $i \in I[1, N]$.

Assumption 4.3 The pairs (X_i, U_i) satisfy the following regulator equations, for $i \in I[1, N]$

$$\begin{cases} A_i X_i + B_i U_i + E_i = X_i A_0 \\ C_i X_i - C_0 = 0. \end{cases} \tag{4.3}$$

Remark 4.1 Assumptions 4.1–4.3 are standard assumptions for studying the output regulation problem for the heterogeneous linear multiagent system such as Cai et al. (2017) and Deng and Wen (2020). Assumption 4.1 is weak, where the subgraph $\mathcal{G}$ does not have to be connected. It is worth emphasizing that the regulator equations (4.3) have solutions if and only if $\operatorname{rank}\left(\begin{bmatrix} A_i - \lambda I & B_i \\ C_i & 0 \end{bmatrix}\right) = n_i + s,\ \lambda \in \sigma(A_0)$.

4.2.2 DoS Attacks

The DoS attacks considered in this chapter occur on edge (j, i) in the communication topology $\bar{\mathcal{G}}$. Denote by $\tilde{t}_m^{ij}$ the moment of the mth attack that occurs on edge (j, i). Use $[\tilde{t}_m^{ij}, \tilde{t}_m^{ij} + \pi_m^{ij})$ to denote the time interval of the mth attack on the edge (j, i), and π_m^{ij} is the attack duration. Use $T_a^{(j,i)}(t_1, t) := [t_1, t] \cap \cup_{m=1}^{\infty}[\tilde{t}_m^{ij}, \tilde{t}_m^{ij} + \pi_m^{ij})$ to represent the set that the attack happens on the edge (j, i). Use $T_a(t_1, t) := [t_1, t] \cap \cup_{(j,i)\in\bar{\mathcal{E}}} T_a^{(j,i)}(t_1, t)$ to denote the set in which at least one edge is attacked in the interval $[t_1, t]$. Use $T_s(t_1, t) := [t_1, t] \backslash T_a(t_1, t)$ to denote the reliable time interval on $[t_1, t]$.

If the edge $(j, i) \in \bar{\mathcal{E}}$ is attacked, the weight of the edge (j, i) is set to $a_{ij}(t) = 0$. Otherwise the weight doesn't change. The corresponding pinning matrix is described by $\mathcal{A}_0(t) = \operatorname{diag}\{a_{10}(t), \ldots, a_{1N}(t)\}$, and $\mathcal{L}(t)$ and $\mathcal{H}(t)$ are defined similarly. It can be obtained that $\lambda_1(\mathcal{H}) > 0$ if $t \in T_s(0, \infty)$ and $\lambda_m(\mathcal{H}(t)) \ge 0$ when $t \in T_a(0, \infty)$. In this case, the following assumption is given to deal with the DoS attack.

Assumption 4.4 There exists $0 < p < 1$ such that $|T_a(t_1, t)| \le p(t - t_1)$ for $t > t_1 > 0$, where $|T_a(t_1, t)|$ denotes the length of $T_a(t_1, t)$.

Remark 4.2 Assumption 4.4 states that the length of the attack time interval $T_a(t_1, t)$ cannot exceed the fraction of the length of the interval $[t_1, t]$. Similar assumptions can also be found in Deng et al. (2022) and Deng and Wen (2020).

4.2.3 Actuator Faults

The lth actuator of the ith agent under the nth fault model is given by

$$u_{il}^{f_n} = \epsilon_{il}^n u_{il},\ l \in I[1, m],\ n \in I[1, N] \tag{4.4}$$

where $u_{il}^{f_n}$ and u_{il} indicate the output and the input for the ith agent, respectively, ϵ_{il}^n indicates the multiplicative factor with $0 \leq \underline{\epsilon}_{il}^n \leq \epsilon_{il}^n \leq \bar{\epsilon}_{il}^n \leq 1$, and $\underline{\epsilon}_{il}^n$ and $\bar{\epsilon}_{il}^n$ are upper and lower bounds of ϵ_{il}^n. It is worth emphasizing that the fault model in Eq. (4.4) includes the fault-free case, ($\underline{\epsilon}_{il}^n = \bar{\epsilon}_{il}^n = 1$), the partial loss-of-effectiveness case ($0 \leq \underline{\epsilon}_{il}^n \leq \bar{\epsilon}_{il}^n \leq 1$), and the outage fault case ($\underline{\epsilon}_{il}^n = \bar{\epsilon}_{il}^n = 0$).

For writing simplicity, the following variables are defined. $u_i^{f_n} = \Lambda_i^n u_i$, $u_i^{f_n} = col(u_{i1}^{fn}, \ldots, u_{in}^{fn})$, $u_i = \text{col}(u_{i1}, \ldots, u_{im})$, $\Lambda_i^n = \text{diag}\{\epsilon_{i1}^n, \ldots, \epsilon_{in}^n\}$, $\Lambda_i = \text{diag}\{\epsilon_{i1}, \ldots, \epsilon_{im}\}$, $\Lambda_i \in \{\Lambda_i^1, \ldots, \Lambda_i^N\}$.

Based on the above definition, we can write the actuator fault model for agent i as

$$u_i^f(t) = \Lambda_i u_i(t),\ i \in I[1, N]. \tag{4.5}$$

Assumption 4.5 The condition of $rank[B_i \Lambda_i] = rank[B_i]$ holds, where Λ_i, $i \in I[1, N]$, are given in Eq. (4.5),

In this case, the following assumption is given to deal with the actuator fault.

Lemma 4.1 (Deng and Wen 2020) *Under the condition of Assumptions 4.2, 4.3, and 4.5, the regulator equations have the solution* (X_i, U_i),

$$\begin{cases} A_i X_i + B_i \Lambda_i U_i + E_i = X_i A_0 \\ C_i X_i - C_0 = 0,\ i \in I[1, N]. \end{cases} \tag{4.6}$$

Lemma 4.2 (Li and Yang 2012) *Under the condition of Assumption 4.5, there exist positive constants* $o_1, \ldots, o_N$ *such that* $B_i \Lambda_i B_i^T \geq o_i B_i B_i^T$ *holds,* $i \in I[1, N]$.

4.3 Main Results

4.3.1 Adaptive Edge-Based Event-Triggered Observers Under DoS Attacks

To estimate the leader matrix and state under DoS attack, the following edge-based event-triggered observer is designed for the ith follower as follows:

$$\dot{G}_i = -\gamma_1 \sum_{j \in I[0,N]} a_{ij}(t)(\tilde{G}_{ij} - \tilde{G}_{ji}) \tag{4.7a}$$

$$\dot{\eta}_i = \bar{A}_i \eta_i - \sum_{j \in I[0,N]} c_{ij} a_{ij}(t) \tilde{\xi}_{ij} \tag{4.7b}$$

where γ_1 is the given positive constant, $G_0 = \mathrm{col}(\mathrm{vec}(A_0), \mathrm{vec}(C_0^T))$, $\bar{A}_i = M_{n_0}^{n_0+s}(G_i)(:, 1:n_0)$, $\tilde{G}_{ij} = G_i(t_k^{ij})$, $\tilde{G}_{i0} = G_i(t_k^{i0})$, $\tilde{G}_{0i} = G_0$, $\tilde{\eta}_{ij} = \eta_i(t_k^{ij})$, $\tilde{\eta}_{i0} = \eta_i(t_k^{i0})$, $\tilde{\xi}_{ij} = \tilde{\eta}_{ij} - \tilde{\eta}_{ji}$ for $j \neq 0$, $\tilde{\xi}_{i0} = \tilde{\eta}_{i0} - x_0$, G_i is the estimate of G_0, η_i is the state of the adaptive edge-based event-triggered observer, c_{ij} and c_{i0} are the adaptive gains, and t_k^{ij}, t_k^{i0} are triggering instants for the edges (j, i) and $(0, i)$, respectively. The adaptive weights c_{ij}, $i \in I[1, N]$, $j \in I[0, N]$, are designed as follows:

$$\dot{c}_{ij} = o_{ij} a_{ij}(t)(-\kappa_{ij} c_{ij} + \|\tilde{\xi}_{ij}\|^2) \tag{4.8}$$

where o_{ij}, κ_{ij} are given positive constants with $\kappa_{ij} = \kappa_{ji}$, $c_{ij}(0) = c_{ji}(0)$.

Define the following measurement errors:

$$e_{ij} = \tilde{G}_{ij} - G_i, \ \text{for } t \in [t_k^{ij}, t_{k+1}^{ij}),$$

$$\bar{e}_{ij} = \tilde{\eta}_{ij} - \eta_i, \ \text{for } t \in [t_k^{ij}, t_{k+1}^{ij}),$$

where $i \in I[1, N]$, $j \in I[0, N]$, and k is the non-negative integer.

Design the following triggering functions:

$$h_{ij} = \|e_{ij}\|^2 - \frac{1}{4}\|\tilde{G}_{ij} - \tilde{G}_{ji}\|^2 - \mu_{ij} e^{-\upsilon_{ij} t} \tag{4.9a}$$

$$h_{i0} = \|e_{i0}\|^2 - \|\tilde{G}_{i0} - G_0\|^2 - \mu_{i0} e^{-\upsilon_{i0} t} \tag{4.9b}$$

and

$$f_{ij} = (1 + 2\iota c_{ij})\|\bar{e}_{ij}\|^2 - \frac{1}{4}\|\tilde{\xi}_{ij}\|^2 - \beta_{ij} e^{-\gamma_{ij} t} \tag{4.10a}$$

$$f_{i0} = \frac{1}{2}(1 + 2\iota c_{i0})\|\bar{e}_{i0}\|^2 - \frac{1}{2}\|\tilde{\xi}_{i0}\|^2 - \beta_{i0}e^{-\gamma_{i0}t} \tag{4.10b}$$

where μ_{ij}, υ_{ij}, ι, β_{ij}, γ_{ij} are given positive constants.

The triggering instant t_k^{ij} is designed as follows:

$$t_{k+1}^{ij} = \inf\{t > t_k^{ij} | h_{ij} \geq 0, \text{ or } f_{ij} \geq 0, \text{ or } \tilde{t}_m^{ij}\}, \tag{4.11}$$

where $i \in I[1, N]$, $j \in I[0, N]$.

Define the following observer errors: $\bar{G}_i = G_i - G_0$, $\bar{\eta}_i = \eta_i - x_0$ for $i \in I[1, N]$. Set $\bar{G}$=col$(\bar{G}_1, \ldots, \bar{G}_N)$, $\bar{\eta}$ = col$(\bar{\eta}_1, \ldots, \bar{\eta}_N)$.

Then, the dynamics of $\bar{G}_i$ and $\bar{\eta}_i$ satisfy

$$\dot{\bar{G}}_i = -\gamma_1 \sum_{j\in I[0,N]} a_{ij}(t)(\tilde{G}_{ij} - \tilde{G}_{ji}) \tag{4.12a}$$

$$\dot{\bar{\eta}}_i = \bar{A}_i\eta_i - A_0x_0 - \sum_{j\in I[0,N]} c_{ij}a_{ij}(t)\tilde{\xi}_{ij} \tag{4.12b}$$

$$\dot{c}_{ij} = o_{ij}a_{ij}(t)(-\kappa_{ij}c_{ij} + \|\tilde{\xi}_{ij}\|^2). \tag{4.12c}$$

Theorem 4.1 *Under the condition of Assumptions 4.1 and 4.2, the errors $\bar{G}_i$ and $\bar{\eta}_i$ for the edge-based event-triggered observers in* (4.7) *can exponentially converge to zero and the bounded set* (4.31)*, respectively, if the following conditions are satisfied, $\frac{\vartheta_1}{\vartheta_1+\beta_1} < p < 1$. In addition, the multiagent system does not exhibit the Zeno behavior under the triggering mechanism composed of Eqs.* (4.9)–(4.11).

Proof Firstly, we show that the proposed observer can estimate the leader's system matrix under DoS attacks. Choose the following Lyapunov function:

$$V_1 = \frac{1}{2} \sum_{i\in I[0,N]} \bar{G}_i^T\bar{G}_i. \tag{4.13}$$

In light of $\tilde{G}_{i0} - G_0 = e_{i0} + \bar{G}_i$, $\tilde{G}_{ij} - \tilde{G}_{ji} = e_{ij} - e_{ji} + \bar{G}_i - \bar{G}_j$, and Lemma 1.6, we calculate the derivative of V_1 along Eq. (4.12) as

$$\begin{aligned}\dot{V}_1 \leq & -\frac{\gamma_1}{2} \sum_{i\in I[1,N]} \sum_{j\in I[1,N]} a_{ij}(t)\bar{G}_i^T(\bar{G}_i - \bar{G}_j) \\ & -\frac{\gamma_1}{4} \sum_{i\in I[1,N]} \sum_{j\in I[1,N]} a_{ij}(t)\|\tilde{G}_{ij} - \tilde{G}_{ji}\|^2\end{aligned}$$

$$+\gamma_1 \sum_{i\in I[1,N]} \sum_{j\in I[1,N]} a_{ij}(t)\|e_{ij}\|^2 - \frac{\gamma_1}{2} \sum_{i\in I[1,N]} a_{i0}(t)\bar{G}_i^T\bar{G}_i$$
$$-\frac{\gamma_1}{2} \sum_{i\in I[1,N]} a_{i0}(t)\|\tilde{G}_{i0} - G_0\|^2 + \frac{\gamma_1}{2} \sum_{i\in I[1,N]} a_{i0}(t)\|e_{i0}\|^2. \tag{4.14}$$

Substituting triggering mechanism into above Eq. (4.14), one obtains that

$$\dot{V}_1 \le -\frac{\gamma_1}{2}\bar{G}^T(\mathcal{H}(t)\otimes I)\bar{G} + \psi_{ij}(t), \tag{4.15}$$

where $\psi_{ij}(t) = \gamma_1 \sum_{i\in I[1,N]} \sum_{j\in I[1,N]} a_{ij}(t)\mu_{ij}e^{-\upsilon_{ij}t} + \frac{\gamma_1}{2}\sum_{i\in I[1,N]} a_{i0}(t)\mu_{i0}e^{-\upsilon_{i0}t}$ is given.

Subsequently, consider the attack interval and the security interval, respectively. If $t \in T_s[0,\infty)$, Eq. (4.15) can be written as

$$\dot{V}_1 \le -\frac{\gamma_1}{2}\lambda_1(\mathcal{H})\bar{G}^T\bar{G} + \psi_{ij}(t) \le -\gamma_1\lambda_1(\mathcal{H})V_1 + \psi_{ij}(t), \tag{4.16}$$

where $\lambda_1(\mathcal{H})$ denotes the smallest eigenvalue of the matrix $\mathcal{H}$.

If $t \in T_a[0,\infty)$, Eq. (4.15) can be written as

$$\dot{V}_1 \le \varepsilon V_1 + \psi_{ij}(t) \tag{4.17}$$

where $\psi_{ij}(t)$ is given in Eq. (4.15), and ε is any small positive constant.

In light of Lemma 1.7, Eqs. (4.16) and (4.17), one obtains that

$$\begin{aligned} V_1 \le & e^{-\gamma_1\lambda_1(\mathcal{H})|T_s(0,t)|}e^{\varepsilon|T_a(0,t)|}V_1(0) + \int_{T_a(0,t)} \psi_{ij}(\tau)e^{\varepsilon|T_a(\tau,t)|-\lambda_1(\mathcal{H})|T_s(\tau,t)|}d\tau \\ & + \int_{T_s(0,t)} \psi_{ij}(\tau)e^{\varepsilon|T_a(\tau,t)|-\lambda_m(\mathcal{H})|T_s(\tau,t)|}d\tau \\ \le & e^{-\vartheta t}V_1(0) + \int_{T_s(0,t)} \psi_{ij}(\tau)e^{-\vartheta t}e^{\vartheta\tau}d\tau + \int_{T_a(0,t)} \psi_{ij}(\tau)e^{-\vartheta t}e^{\vartheta\tau}d\tau \\ \le & e^{-\vartheta t}V_1(0) + \int_{T_s(0,t)} \psi_{ij}(\tau)d\tau + \int_{T_a(0,t)} \psi_{ij}(\tau)d\tau \end{aligned} \tag{4.18}$$

where $\vartheta = \gamma_1\lambda_m(\mathcal{H})(1-p) - \varepsilon p$ is defined. It is worth emphasizing that the parameter ϑ can be ensured to be positive by choosing a sufficiently small parameter ε.

Conclusively, based on Eq. (4.18), one can obtain that error $\bar{G}_i$ exponentially converges to zero at a rate of ϑ. Therefore, the proposed observer in Eq. (4.7a) can successfully estimate the leader's matrix under the influence of DoS attack.

Secondly, we show that the proposed observer (4.7b) can estimate the leader's state x_0 under DoS attacks. Choose the following Lyapunov-like function:

$$V_2 = \frac{1}{2} \sum_{i \in I[1,N]} \bar{\eta}_i^T \bar{\eta}_i + \sum_{i \in I[1,N]} \sum_{j \in I[1,N]} \frac{\bar{c}_{ij}^2}{8 o_{ij}} + \sum_{i \in I[1,N]} \frac{\bar{c}_{i0}^2}{4 o_{i0}} \tag{4.19}$$

where $\bar{c}_{ij} = c_{ij} - \alpha$, α is a positive constant to be specified, and o_{ij} and o_{i0} are given in Eq. (4.8).

Calculate the derivative of V_2 along (4.12b) and (4.12c)

$$\begin{aligned} \dot{V}_2 = & \sum_{i \in I[1,N]} \bar{\eta}_i^T \big(\bar{A}_i \eta_i - A_0 x_0 - \sum_{j \in I[0,N]} c_{ij} a_{ij}(t) \tilde{\xi}_{ij}\big) \\ & + \sum_{i \in I[1,N]} \sum_{j \in I[1,N]} \frac{\bar{c}_{ij}}{4 o_{ij}} \dot{c}_{ij} + \sum_{i \in I[1,N]} \frac{\bar{c}_{i0}}{2 o_{ij}} \dot{c}_{i0}. \end{aligned} \tag{4.20}$$

According to Lemma 1.6 and the definition of $\bar{\eta}_i = \eta_i - x_0$, we can obtain the following inequality:

$$\begin{aligned} & \sum_{i \in I[1,N]} \bar{\eta}_i^T (\bar{A}_i \eta_i - A_0 x_0) \\ = & \sum_{i \in I[1,N]} \bar{\eta}_i^T \big((\bar{A}_i - A_0)(\bar{\eta}_i + x_0) + A_0 \bar{\eta}_i\big) \\ \le & \bar{\eta}^T \bar{\eta} + \bar{\eta}^T (I \otimes A_0) \bar{\eta} + \frac{1}{2} \sum_{i \in I[1,N]} \|(\bar{A}_i - A_0) \bar{\eta}_i\|^2 \\ & + \frac{1}{2} \sum_{i \in I[1,N]} \|(\bar{A}_i - A_0) x_0\|^2. \end{aligned} \tag{4.21}$$

Thanks to Lemma 1.6, $\tilde{\xi}_{ij} = \bar{e}_{ij} - \bar{e}_{ji} + \bar{\eta}_i - \bar{\eta}_j$, $\tilde{\xi}_{i0} = \bar{e}_{i0} + \bar{\eta}_i$, we can obtain the following inequality:

$$\begin{aligned} & - \sum_{i \in I[1,N]} \sum_{j \in I[0,N]} c_{ij} a_{ij}(t) \bar{\eta}_i^T \tilde{\xi}_{ij} \\ \le & - \frac{1}{2} \sum_{i \in I[1,N]} \sum_{j \in I[0,N]} c_{ij} a_{ij}(t) \bar{\eta}_i^T (\bar{\eta}_i - \bar{\eta}_j) \\ & - \frac{1}{2} \sum_{i \in I[1,N]} c_{i0} a_{i0}(t) \bar{\eta}_i^T \bar{\eta}_i + \sum_{i \in I[1,N]} \sum_{j \in I[1,N]} c_{ij} a_{ij}(t) \bar{e}_{ij}^T \bar{e}_{ij} \\ & - \frac{1}{4} \sum_{i \in I[1,N]} \sum_{j \in I[1,N]} c_{ij} a_{ij}(t) \tilde{\xi}_{ij}^T \tilde{\xi}_{ij} \\ & + \frac{1}{2} \sum_{i \in I[1,N]} c_{i0} a_{i0}(t) \bar{e}_{i0}^T \bar{e}_{i0} - \frac{1}{2} \sum_{i \in I[1,N]} c_{i0} a_{i0}(t) \tilde{\xi}_{i0}^T \tilde{\xi}_{i0}. \end{aligned} \tag{4.22}$$

Invoking Lemma 1.6 and Eq. (4.12c), we can obtain the following inequality:

$$
\begin{aligned}
&\sum_{i\in I[1,N]}\sum_{j\in I[1,N]}\frac{\bar{c}_{ij}}{4o_{ij}}\dot{c}_{ij}+\sum_{i\in I[1,N]}\frac{\bar{c}_{i0}}{2o_{i0}}\dot{c}_{i0}\\
\leq&\frac{1}{8}\sum_{i\in I[1,N]}\sum_{j\in I[1,N]}\kappa_{ij}a_{ij}(t)(\alpha^2-\bar{c}_{ij}^2)+\frac{1}{4}\sum_{i\in I[1,N]}\kappa_{i0}a_{i0}(t)(\alpha^2-\bar{c}_{i0}^2)\\
&+\sum_{i\in I[1,N]}\sum_{j\in I[1,N]}\frac{c_{ij}}{4}a_{ij}(t)\|\tilde{\xi}_{ij}\|^2+\sum_{i\in I[1,N]}\frac{c_{i0}}{2}a_{i0}(t)\|\tilde{\xi}_{i0}\|^2\\
&-\sum_{i\in I[1,N]}\sum_{j\in I[1,N]}\frac{\alpha}{8}a_{ij}(t)\|\tilde{\xi}_{ij}\|^2-\sum_{i\in I[1,N]}\frac{\alpha}{4}a_{i0}(t)\|\tilde{\xi}_{i0}\|^2\\
&+\frac{\alpha}{2}\sum_{i\in I[1,N]}\sum_{j\in I[1,N]}a_{ij}(t)\|\bar{e}_{ij}\|^2+\frac{\alpha}{4}\sum_{i\in I[1,N]}a_{i0}(t)\|\bar{e}_{i0}\|^2\\
&-\frac{\alpha}{8}\sum_{i\in I[1,N]}\sum_{j\in I[1,N]}a_{ij}(t)\bar{\eta}_i^T(\bar{\eta}_i-\bar{\eta}_j)\\
&-\frac{\alpha}{8}\sum_{i\in I[1,N]}a_{i0}(t)\bar{\eta}_i^T\bar{\eta}_i. \qquad (4.23)
\end{aligned}
$$

Based on the above analysis, we can obtain that $\|\bar{A}_i-A_0\|^2$ is bounded for $i=1,\ldots,N$. Therefore, we can obtain that $\|\bar{A}_i-A_0\|^2\leq\delta$, where δ is the finite positive constant.

Substituting Eqs. (4.20)–(4.23) into Eq. (4.19), we can obtain the following inequality:

$$
\begin{aligned}
\dot{V}_2\leq&\frac{1}{2}\bar{\eta}^T\big(I_N\otimes(A_0+A_0^T+2I+\delta I)-\frac{\alpha}{4}\mathcal{H}(t)\otimes I\big)\bar{\eta}\\
&-\frac{1}{8}\sum_{i\in I[1,N]}\sum_{j\in I[1,N]}\kappa_{ij}a_{ij}(t)\bar{c}_{ij}^2-\frac{1}{4}\sum_{i\in I[1,N]}\kappa_{i0}a_{i0}(t)\bar{c}_{i0}^2+\delta_1\\
&+\frac{\alpha}{2}\sum_{i\in I[1,N]}\sum_{j\in I[1,N]}a_{ij}(t)\big((1+2\iota c_{ij})\|\bar{e}_{ij}\|^2-\frac{1}{4}\|\tilde{\xi}_{ij}\|^2\big)\\
&+\frac{\alpha}{2}\sum_{i\in I[1,N]}a_{i0}(t)\big((\frac{1}{2}+\iota c_{i0})\|\bar{e}_{i0}\|^2-\frac{1}{2}\|\tilde{\xi}_{i0}\|^2\big)\\
&+\frac{1}{2}\sum_{i\in I[1,N]}\|(\bar{A}_i-A_0)x_0\|^2 \qquad (4.24)
\end{aligned}
$$

where $\delta_1=\frac{1}{8}\sum_{i\in I[1,N]}\sum_{j\in I[1,N]}\kappa_{ij}a_{ij}(t)\alpha^2+\frac{1}{4}\sum_{i\in I[1,N]}\kappa_{i0}a_{i0}(t)\alpha^2$ and $\alpha>\frac{1}{\iota}$ are defined.

According to the triggering mechanism in Eq. (4.11) and triggering functions (4.10a) and (4.10b), we can obtain that Eq. (4.23) satisfies

$$\dot{V}_2 \leq \frac{1}{2}\bar{\eta}^T \left(I_N \otimes (A_0 + A_0^T + 2I + \delta I) - \frac{\alpha}{4}\mathcal{H}(t) \otimes I\right)\bar{\eta}$$
$$- \frac{1}{8}\sum_{i\in I[1,N]}\sum_{j\in I[1,N]} \kappa_{ij}a_{ij}(t)\bar{c}_{ij}^2$$
$$- \frac{1}{4}\sum_{i\in I[1,N]} \kappa_{i0}a_{i0}(t)\bar{c}_{i0}^2 + \delta_1$$
$$+ \frac{1}{2}\sum_{i\in I[1,N]} \|(\bar{A}_i - A_0)x_0\|^2 + \bar{\psi}_{ij}(t) \tag{4.25}$$

where $\bar{\psi}_{ij}(t) = \frac{\alpha}{2}\sum_{i\in I[1,N]}\sum_{j\in I[1,N]} a_{ij}(t)\beta_{ij}e^{-\gamma_{ij}t} + \frac{\alpha}{2}\sum_{i\in I[1,N]} a_{i0}(t)\beta_{i0}e^{-\gamma_{i0}t}$ is defined. It is not difficult to obtain that $\lim_{t\to\infty}\bar{\psi}_{ij}(t) = 0$ and $\lim_{t\to\infty}\frac{1}{2}\sum_{i\in I[1,N]}\|(\bar{A}_i - A_0)x_0\|^2 = 0$. Therefore, one obtains that there exists $\delta_2 > 0$ such that $\frac{1}{2}\sum_{i\in I[1,N]}\|(\bar{A}_i - A_0)x_0\|^2 \leq \delta_2$ for $t > T_1$.

Subsequently, consider the attack interval and security, respectively. The following symbol definition is given before proceeding, $\alpha > \max\{\frac{1}{\iota}, \frac{2\varpi\sqrt{M}}{\lambda_1(\mathcal{H})}\}$.

If $t \in T_s[0,\infty)$, we can obtain that $\lambda_1(\mathcal{H})$ is strictly positive, so we obtain that there exists the positive constant $\beta > 0$ such that $A_0 + A_0^T + 2I + \delta I - \frac{\alpha}{4}\lambda_1(\mathcal{H}) \leq -\beta I$. Obviously, Eq. (4.25) can be converted into

$$\dot{V}_2 \leq -\frac{\beta}{2}\bar{\eta}^T\bar{\eta} - \frac{1}{8}\sum_{i\in I[1,N]}\sum_{j\in I[1,N]} \kappa_{ij}a_{ij}(t)\bar{c}_{ij}^2$$
$$- \frac{1}{4}\sum_{i\in I[1,N]} \kappa_{i0}a_{i0}(t)\bar{c}_{i0}^2$$
$$+ \delta_1 + \bar{\psi}_{ij}(t) + \frac{1}{2}\sum_{i\in I[1,N]} \|(\bar{A}_i - A_0)x_0\|^2$$
$$\leq -\vartheta_1 V_2 - \frac{1}{2}(\beta - \vartheta_1)\bar{\eta}^T\bar{\eta} + \delta_1 + \delta_2 + \bar{\psi}_{ij}(t) \tag{4.26}$$

where $0 < \vartheta_1 < \min_{(j,i)\in\bar{\mathcal{E}}}\{\kappa_{ij}o_{ij}|a_{ij}|, \beta\}$ is defined.

If $t \in T_a[0,\infty)$, there exist the positive constants $\beta_1 > 0$ and $\beta_2 > 0$ such that $A_0 + A_0^T + 2I + \delta I - \beta_1 I < -\beta_2 I$. Obviously, Eq. (4.26) can be converted into

$$\dot{V}_2 \leq \beta_1 V_2 - \frac{\beta_2}{2}\bar{\eta}^T\bar{\eta} + \delta_1 + \delta_2 + \bar{\psi}_{ij}(t). \tag{4.27}$$

Select ϱ_1 such that $\varrho_1 = \min\{\beta - \vartheta_1, \frac{\beta_2}{2}\}$. When $\|\bar{\eta}\| > \sqrt{2(\delta_1 + \delta_2)/\varrho_1}$, Eqs. (4.26) and (4.27) can be respectively converted into

$$\dot{V}_2 \leq \vartheta_1 V_2 + \bar{\psi}_{ij}(t), \quad \text{for } t \in T_s(0,\infty) \tag{4.28}$$

and

$$\dot{V}_2 \leq \beta_1 V_2 + \bar{\psi}_{ij}(t), \ \text{for}\ t \in T_a(0, \infty). \tag{4.29}$$

In light of Lemma 1.7, Eqs. (4.28) and (4.30), one can obtain that

$$\begin{aligned} V_2 \leq & e^{-\vartheta_1 |T_s(0,t)|} e^{\beta_1 |T_a(0,t)|} V_2(0) + \int_{T_s(0,t)} \bar{\psi}_{ij}(\tau) e^{\beta_1 |T_a(\tau,t)| - \vartheta_1 |T_s(\tau,t)|} d\tau \\ & + \int_{T_a(0,t)} \bar{\psi}_{ij}(\tau) e^{\beta_1 |T_a(\tau,t)| - \vartheta_1 |T_s(\tau,t)|} d\tau \\ \leq & e^{-\vartheta_2 t} V_2(0) + \int_{T_s(0,t)} \bar{\psi}_{ij}(\tau) e^{-(t-\tau)\vartheta_2} d\tau + \int_{T_a(0,t)} \bar{\psi}_{ij}(\tau) e^{-(t-\tau)\vartheta_2} d\tau \end{aligned} \tag{4.30}$$

where $\vartheta_2 = \vartheta_1(1-p) - \beta_1 p$ is positive from the condition of Theorem 4.1.

Accordingly, according to Eq. (4.30), we can obtain that $\bar{\eta}$ converges exponentially to the bounded set,

$$\Theta = \{\bar{\eta} \mid \|\bar{\eta}\| \leq \sqrt{\frac{2(\delta_1 + \delta_2)}{\varrho_1}}\}, \tag{4.31}$$

i.e., the proposed observer can successfully estimate the leader's state under the influence of DoS.

Finally, we prove that the multiagent system composed of Eqs. (4.1) and (4.2) does not exhibit the Zeno behavior under the proposed triggering mechanism in Eq. (4.11). We adopt the reduction to absurdity here. It may be assumed that Zeno behavior occurs on the edge (j, i), $i \in I[1, N]$, $j \in I[0, N]$, which indicates that $\lim_{k' \to \infty} t^{ij}_{k'+1} = t^{ij}_{\infty} < \infty$.

Consider the adjacent inter-event interval $[t^{ij}_{k'}, t^{ij}_{k'+1})$. When $t = t^{ij}_{k'+1}$, we have $e_{ij}(t^{ij}_{k'+1}) = 0$, $\bar{e}_{ij}(t^{ij}_{k'+1}) = 0$, and $h_{ij}(t^{ij}_{k'+1}) \geq 0$ or $f_{ij}(t^{ij}_{k'+1}) \geq 0$. According to triggering functions in Eqs. (4.9) and (4.10),

ignoring the state-dependent terms, and combining the triggering mechanism in Eq. (4.11), we can obtain the following more conservative condition:

$$\|e_{ij}(t^{ij}_{k'+1})\|^2 \geq \mu_{ij} e^{-\upsilon_{ij} t^{ij}_{k'+1}},$$

$$(1 + 2\iota c_{ij}) \|\bar{e}_{ij}(t^{ij}_{k'+1})\|^2 \geq \beta_{ij} e^{-\gamma_{ij} t^{ij}_{k'+1}}.$$

Therefore, one can obtain that $0 = \|e_{ij}(t^{ij}_{\infty})\|^2 \geq \mu_{ij} e^{-\upsilon_{ij} t^{ij}_{\infty}} > 0$ or $0 = (1 + 2\iota c_{ij}) \|\bar{e}_{ij}(t^{ij}_{\infty})\|^2 \geq \beta_{ij} e^{-\gamma_{ij} t^{ij}_{\infty}} > 0$. Obviously, that's not true. Therefore, the assumption is wrong, and Zeno behavior does not occur at the edge (j, i) under the triggering mechanism in Eq. (4.11). In summary, the multiagent system composed of Eqs. (4.1) and (4.2) does not exhibit the Zeno behavior under the proposed trigger mechanism in (4.11). The proof is completed. □

Remark 4.3 Only well-informed followers have access to the leader's system matrix, for which the observer in Eq. (4.7a) is designed to estimate the leader's system matrix, and accordingly, time-varying matrix is introduced into the design of the observer in Eq. (4.7b) for estimating the leader's state, for which the means of dealing with time-varying matrices are still scarce and challenging. Note that the current methods of dealing with time-varying matrices are still few and challenging. The methods in Cai et al. (2017), Deng and Wen (2020), Deng et al. (2022), and Qian et al. (2021) can handle time-varying matrices, but they have the disadvantage of relying on global information about the communication topology, and although method in Qian et al. (2021) can handle time-varying matrices in a fully distributed manner, it is not applicable to situations where DoS attacks are present. In order to deal with time-varying matrices under the influence of DoS attacks in a fully distributed manner, this chapter introduces the crucial step, i.e., formula (4.21). Further we analyze to obtain that $\|\bar{A}_i - A_0\|$ is bounded and $\lim_{t\to\infty} \|(\bar{A}_i - A_0)x_0\|^2 = 0$.

Remark 4.4 Under the influence of the DoS attack, the eigenvalues of the leader-follower matrix $\mathcal{H}(t)$ are the non-negative, and the matrix $\mathcal{H}(t)$ is symmetric. Therefore, Eqs. (4.15) and (4.25) need to be simplified into two scenarios, namely the safe interval and the attack interval.

Remark 4.5 In the presence of DoS attack, Theorem 3.1 shows that the observer errors converge exponentially to 0 and a small adjustable bounded set, respectively, and that the multiagent system does not exhibit the Zeno behavior.

4.3.2 Adaptive Event-Triggered Fault-Tolerant Controller

This chapter considers the more general case where only well-informed followers have direct access to the leader system matrix and state. For heterogeneous multiagent systems, to realize the output regulation problem, the design of the controller depends on the solution of the regulator equations, which depend on the leader system matrix. Due to the unknown leader system matrix, there exists the time-varying matrix in the regulator equation, and in order to obtain the solution of the regulator equation, the following lemma is presented.

Lemma 4.3 (Cai et al. 2017) *Under the condition of Assumptions 4.2 and 4.3, the following equation*

$$\dot{\zeta}_i = -\varrho \Pi_i^T (\Pi_i \zeta_i - \Xi_i) \tag{4.32}$$

has a unique bounded solution for any given initial state $\zeta_i(0)$*, where* $\Pi_i = \bar{A}_i^T \otimes \begin{bmatrix} I_{n_i} & 0_{n_i \times m_i} \\ 0_{s \times n_i} & 0_{s \times m_i} \end{bmatrix} - I_{n_0} \otimes \begin{bmatrix} A_i & B_i \\ C_i & 0_{s \times m_i} \end{bmatrix}$, $\Xi_i = vec\left(\begin{bmatrix} E_i \\ -\bar{C}_i \end{bmatrix}\right)$ *with* $\bar{C}_i = \left(M_{n_0}^{n_0+s}(G_i)\right.$ $\left.(:, n_0+1 : n_0+s)\right)^T$ *are given, and* ϱ *is the positive constant. If* $\bar{G}_i$ *exponentially converges to zero, one can obtain that* $M_{n_i+m_i}^{n_0}(\zeta_i(t)) = \begin{bmatrix} X_i(t) \\ \Lambda_i U_i(t) \end{bmatrix}$ *can exponentially converge to* $\begin{bmatrix} X_i^* \\ \Lambda_i U_i^* \end{bmatrix}$*, where* (X_i^*, U_i^*)*,* $i \in I[1, N]$ *are given in Eq. (4.6).*

The adaptive event-triggered fault-tolerant controller is designed to compensate for the effect of actuator fault on system stability and to reduce the communication frequency from the controller to the actuator to realize the output regulation problem of the multiagent system. The controller designed for the ith follower is

$$\begin{aligned} u_i &= \bar{W}_{1i}(t_h^i)\big(\bar{x}_i(t_h^i) - X_i(t_h^i)\eta_i(t_h^i)\big) + \bar{W}_{2i}(t_h^i)\eta_i(t_h^i) \\ \dot{\bar{x}}_i &= A_i\bar{x}_i + B_i u_i^f + E_i\eta_i + \bar{J}_i(C_i\bar{x}_i - y_i) \end{aligned} \tag{4.33}$$

where $\bar{x}_i \in R^{n_i}$ denotes the observer state for follower i, $\bar{W}_{1i} = \mathrm{col}(\bar{W}_{1i}^1, \ldots, \bar{W}_{1i}^{m_i}) \in R^{m_i \times n_i}$ with $\bar{W}_{1i}^j \in R^{1 \times n_i}$ for $j \in I[1, m_i]$, $\bar{W}_{2i} = \mathrm{col}(\bar{W}_{2i}^1, \ldots, \bar{W}_{2i}^{m_i}) \in R^{m_i \times n_0}$ with $\bar{W}_{2i}^j \in R^{1 \times n_0}$ for $j \in I[1, m_i]$ denote the adaptive feedback matrices to be designed, and $\bar{J}_i \in R^{n_i \times s}$ denotes the output feedback matrix and satisfies $A_i + \bar{J}_i C_i$ is the Hurwitz matrix.

Denote the observer error in Eq. (4.33) as $e_{x_i} = \bar{x}_i - x_i$. Calculate the derivative of $e_{x_i} = \bar{x}_i - x_i$ as

$$\dot{e}_{x_i} = (A_i + \bar{J}_i C_i)e_{x_i} + E_i\bar{\eta}_i. \tag{4.34}$$

Since $\bar{\eta}_i$ converges to the small bounded set according to Theorem 3.1, and the fact that $A_i + \bar{J}_i C_i$ is the Hurwitz matrix, one can obtain that e_{x_i} can converge to any small bounded set by Eq. (4.34). Set $e_x = \mathrm{col}(e_{x_1}, \ldots, e_{x_M})$. It may be assumed that there exists the positive constant δ_3 such that $\|e_x\| \le \delta_3$ for any $t > T_2$.

Set $\tilde{x}_i = \bar{x}_i - X_i(t)\eta_i$. Design the adaptive update rate for $\bar{W}_{1i}^j$ and $\bar{W}_{2i}^j$ as

$$\begin{aligned} \dot{\bar{W}}_{1i}^j &= -\tilde{x}_i^T P_i B_i^j \tilde{x}_i^T \\ \dot{\bar{W}}_{2i}^j &= -\tilde{x}_i^T P_i B_i^j \eta_i^T \end{aligned} \tag{4.35}$$

where η_i is given in Eq. (4.7), $B_i^j \in R^{n_i}$ denotes the jth column for $j \in I[1, m_i]$, and P_i denotes the positive definite matrix to be given.

Remark 4.6 The advantages of the adaptive event-triggered fault-tolerant controller designed in this chapter are summarized in two aspects: (1) The chatter phenomenon can be avoided through the introduction of adaptive feedback matrices. (2) The frequency of the communication from the controller to the actuator can be reduced through the design of the event-triggered mechanism.

To construct the event-triggered mechanism, first define the measurement error

$$e_{u_i} = \bar{W}_{1i}(t_h^i)\tilde{x}_i(t_h^i) - \bar{W}_{1i}\tilde{x}_i + \bar{W}_{2i}(t_h^i)\eta_i(t_h^i) - \bar{W}_{2i}\eta_i$$

where $t \in [t_h^i, t_{h+1}^i)$ is given, and h denotes the hth trigger. Then, construct the following triggering mechanism:

$$t_{h+1}^i = \inf\{t_h^i | f_i^u \geq 0\} \tag{4.36}$$

where f_i^u is the triggering function. The triggering function f_i^u is constructed as

$$f_i^u = \|e_{u_i}\|^2 - \iota_1\|\tilde{x}_i\|^2 - \bar{\beta}_i e^{-\bar{\gamma}_i t} \tag{4.37}$$

where the parameters $\iota_1,\ \bar{\beta}_i, \bar{\gamma}_i$ are positive with $\iota_1 < 2\lambda_1(Q)$, $Q = \text{diag}\{Q_1, \ldots, Q_M\}$ is to be given, and Q_i is the positive definite matrix for $i \in I[1, N]$. Note that $\lambda_1(Q)$ denotes the smallest eigenvalue of the matrix Q.

For subsequent writing simplicity, some variable definitions are given.

$$W_{1i} = -\frac{2}{o_i}B_i^T P_i,\ W_{2i} = U_i,$$

where $o_i,\ P_i,\ U_i$ are given in Lemma 4.2, (4.35), and (4.6), respectively.

$$\bar{A} = \text{diag}\{A_1, \ldots, A_N\}, \bar{A}_0 = I_M \otimes A_0, \bar{B} = \text{diag}\{B_1, \ldots, B_N\},$$

$$\bar{C} = \text{diag}\{C_1, \ldots, C_N\}, \bar{J} = \text{diag}\{\bar{J}_1 \ldots, \bar{J}_N\},$$

$$\bar{\Lambda} = \text{diag}\{\Lambda_1, \ldots, \Lambda_N\}, W_1 = \text{diag}\{W_{11}, \ldots, W_{1N}\},$$

$$\bar{W}_1 = \text{diag}\{\bar{W}_{11} \ldots, \bar{W}_{1N}\}, W_2 = \text{diag}\{W_{21}, \ldots, W_{2N}\},$$

$$\bar{W}_2 = \text{diag}\{\bar{W}_{21} \ldots, \bar{W}_{2N}\}, \bar{X}(t) = \text{diag}\{\bar{X}_1(t), \ldots, \bar{X}_N\},$$

$$\bar{X} = \text{diag}\{X_1, \ldots, X_N\}, \tilde{x} = \text{col}(\tilde{x}_1, \ldots, \tilde{x}_N),$$

$$\bar{\eta} = \eta - \bar{x}_0 \text{ with } \bar{x}_0 = 1_N \otimes x_0.$$

Based on the above definition, we can obtain the derivate of $\tilde{x}$ satisfies

$$\begin{aligned}\dot{\tilde{x}} =& (\bar{A} + \bar{B}\bar{\Lambda}W_1)\tilde{x} + \bar{B}\bar{\Lambda}\big(\bar{W}_1(t_h)\tilde{x}(t_h) - \bar{W}_1\tilde{x}\big) + \bar{B}\bar{\Lambda}(\bar{W}_2(t_h)\eta(t_h) - \bar{W}_2\eta) \\ &+ \bar{B}\bar{\Lambda}(\bar{W}_1 - W_1)\tilde{x} + \bar{B}\bar{\Lambda}(\tilde{W}_2 - W_2)\eta + \Upsilon\end{aligned} \tag{4.38}$$

where $\Upsilon = \bar{X}(t)\bar{A}_0\bar{\eta} - \bar{X}(t)\dot{\bar{\eta}} - \dot{\bar{X}}(t)\eta + (\bar{X}(t) - \bar{X})\bar{A}_0(\mathcal{S} \otimes I)\bar{x}_0 - \bar{A}(\bar{X} - \bar{X}(t))\eta + \bar{J}\bar{C}e_x$ is given.

It follows from Lemma 4.3 that $\dot{\bar{X}}(t)\eta$ can converge to zero. Furthermore, we can obtain that there exists the positive constant τ such that $\|\Upsilon\| \leq \tau\delta_3$ for $t > T_3$.

The following theorem is presented to illustrate the theoretical feasibility of the adaptive event-triggered fault-tolerant controller designed in this chapter.

Theorem 4.2 *Under the conditions of Assumptions 4.1–4.5 and Theorem 4.1, consider the multiagent system composed of Eqs.* (4.1) *and* (4.2), *the proposed adaptive event-triggered fault-tolerant controller in Eq.* (4.33) *and the edge-based event-triggered observer in* (4.7) *can deal with the cooperative output regulation problem if selected matrix P_i satisfies*

$$A_i^T P_i + P_i A_i + \rho_i^{-1} P_i^2 - 2P_i B_i B_i^T P_i + Q_i \leq 0 \tag{4.39}$$

where Q_i is any given positive definite matrix, and parameter ρ_i is positive. In addition, the multiagent system does not exhibit the Zeno behavior under the proposed triggering mechanism in Eq. (4.36).

Proof Consider the following Lyapunov function:

$$V_3 = \sum_{i\in I[1,N]} \tilde{x}_i^T P_i \tilde{x}_i + \sum_{i\in I[1,N]} \sum_{j\in I[1,m_i]} \epsilon_{ij}\big(\hat{W}_{1i}^j(\hat{W}_{1i}^j)^T + \hat{W}_{2i}(\hat{W}_{2i}^j)^T\big)$$

where $\hat{W}_{1i} = \bar{W}_{1i}^j - W_{1i}^j$ and $\hat{W}_{2i} = \bar{W}_{2i}^j - W_{2i}^j$ are given, and W_{1i}^j and W_{2i}^j denote the jth row of matrices W_{1i} and W_{2i}, respectively.

According to Eqs. (4.35) and (4.38), the derivative of V_3 can be obtained that

$$\begin{aligned}
\dot{V}_3 =& \tilde{x}^T(\bar{P}\bar{A} + \bar{A}^T\bar{P} + \bar{P}\bar{B}\bar{\Lambda}W_1 + W_1^T\bar{\Lambda}\bar{B}^T\bar{P})\tilde{x} \\
&+ 2\tilde{x}^T\bar{P}\bar{B}\bar{\Lambda}\big(\bar{W}_1(t_h)\tilde{x}(t_h) - \bar{W}_1\tilde{x} + \bar{W}_2(t_h)\eta(t_h) - \bar{W}_2\eta\big) \\
&+ 2\tilde{x}^T\bar{P}\Upsilon \\
\leq& \tilde{x}^T(\bar{P}\bar{A} + \bar{A}^T\bar{P} - 4o^{-1}\bar{P}\bar{B}\bar{\Lambda}\bar{B}^T\bar{P})\tilde{x} + \frac{1}{2}\sum_{i\in I[1,N]}\|e_{u_i}\|^2 \\
&+ 2\|\tilde{x}^T\bar{P}\bar{B}\Lambda\|^2 + \tilde{x}^T\bar{P}\bar{\rho}^{-1}\bar{P}\tilde{x} + \Upsilon^T\bar{\rho}\Upsilon
\end{aligned} \tag{4.40}$$

where $\bar{P} = I_M \otimes P$ is given, $o = \mathrm{diag}\{o_1, \ldots, o_N\}$, and $\bar{\rho} = \mathrm{diag}\{\rho_1, \ldots, \rho_N\}$.

According to Eq. (4.37), one can obtain that $\delta_4 = \lambda_1(\bar{Q}) - \frac{\iota_1}{2} > 0$. In light of Lemma 4.2 and triggering mechanism in Eq. (4.36), one can obtain that Eq. (4.40) can be written as

$$\dot{V}_3 \leq -\delta_4\tilde{x}^T\tilde{x} + \frac{1}{2}M\bar{\beta}e^{-\bar{\gamma}t} + \Upsilon^T\bar{\rho}\Upsilon \tag{4.41}$$

which indicates that error $\bar{x}$ converges to the adjustable bounded set $\Delta =: \{\bar{x} \mid \|\bar{x}\| \leq \frac{\rho_M\tau^2\delta_3^2}{\delta_4}\}$ with $\rho_M = \max_{\{i\in I[1,N]\}}\{\rho_i\}$.

Denote the output regulation error as $\bar{y}_i = y_i - y_0$. Then, according to system dynamics in Eqs. (4.1) and (4.2), one can obtain that

$$\begin{aligned}\bar{y}_i &= C_i\big(\tilde{x}_i + X_i(t)\eta_i\big) - C_i e_{x_i} - C_0 x_0 \\ &= C_i\tilde{x}_i - C_i e_{x_i} + C_0\bar{\eta}_i \end{aligned} \tag{4.42}$$

where Lemma 4.1 is adopted to obtain the above inequality. Due to the fact that $\|e_x\| \le \delta_3$, $\|\tilde{x}\| \le \frac{\rho_M \tau^2 \delta_3^2}{\delta_4}$, $\|\bar{\eta}\| \le \sqrt{\frac{2(\delta_1+\delta_2)}{\varrho_1}}$ for sufficiently large t, we conclude $\lim\ \sup_{t\to\infty}\|y_i - y_0\| \le \tau_1\varepsilon$ with $\tau_1 = \max_{\{i\in I[1,N]\}}\{\|\bar{C}_i\|, \|C_0\|\}$ and $\varepsilon = \max\{\delta_3, \frac{\rho_M \tau^2 \delta_3^2}{\delta_4}, \sqrt{\frac{2(\delta_1+\delta_2)}{\varrho_1}}\}$. Therefore, the cooperative output regulation problem is solved. Finally, the multi-intelligent systems considered in our analysis do not exhibit Zeno behavior under the proposed triggering mechanism. Note that its proof is similar to Theorem 4.1 and is omitted here. This completes the proof. □

Remark 4.7 Unlike some existing studies on the output regulation problem of heterogeneous linear multiagent systems in Qian et al. (2021), Cai et al. (2017), Deng and Wen (2020), and Zhang et al. (2023a), the approach in this chapter can realize not only asynchronous event-triggered communication of neighbors and event-triggered update of the controller. Therefore, the method in this chapter can effectively save the communication resources.

4.4 Simulation Examples

In this section, we give two examples to argue for the validity and superiority of the methods in this chapter.

Example 4.1 Consider four carts with inverted pendulums as in Deng and Wen (2020). The i $(i = 1, 2, 3, 4)$th cart with inverted pendulum is modeled by

$$\begin{aligned} M_i\ddot{d}_i &= -m_i g\vartheta_i - h_i\dot{d}_i + u_i + \alpha_{i1}x_{01} \\ M_i\iota_i\ddot{\vartheta}_i &= (m_i + M_i)g\vartheta_i + h_i\dot{d}_i - u_i + \alpha_{i2}x_{01} \end{aligned}$$

where ϑ_i expresses the angle of the pendulum, d_i expresses the distance of the cart, u_i expresses the horizontal force applied to the cart, $h_i = 0.15$, $g = 9.81$, $m_i = 0.1 * i$, $M_i = 1 * i$, $\alpha_{i1} = \alpha_{i2} = i$ for $i = 1, 2, 3, 4$. The leader signal is given by $\dot{x}_{01} = x_{02}$, $\dot{x}_{02} = -x_{01}$, where $x_0 = \mathrm{col}(x_{01}, x_{02})$. The leader-follower matrix $\mathcal{H}$ of the communication topology $\bar{\mathcal{G}}$ is modeled by

$$\mathcal{H} = \begin{bmatrix} 2 & 0 & -1 & 0 \\ 0 & 1 & 0 & -1 \\ -1 & 0 & 2 & -1 \\ 0 & -1 & -1 & 2 \end{bmatrix}.$$

In order to apply the methodology of this chapter, the system under consideration is converted into the form shown in Eq. (4.1). Using coordinate transformations, $x_{i1} = d_i + \iota_i \vartheta_i,\ x_{i2} = \dot{x}_{i1},\ x_{i3} = \vartheta_i,\ x_{i4} = \dot{\vartheta}_i$. The dynamics of the cart can be changed into the form of Eq. (4.1), where $A_i = \begin{bmatrix} 0 & 1 & 0 & 0 \\ 0 & 0 & g & 0 \\ 0 & 0 & 0 & 1 \\ 0 & \varpi_{i1} & \varpi_{i2} & \varpi_{i3} \end{bmatrix}$,

$B_i = \begin{bmatrix} 0 \\ 0 \\ b_i \end{bmatrix}$, $C_i = \begin{bmatrix} 1 \\ 0 \\ -\iota_1 \\ 0 \end{bmatrix}^T$, $E_i = \begin{bmatrix} 0 & 0 \\ p_{i1} & 0 \\ 0 & 0 \\ p_{i2} & 0 \end{bmatrix}$, $A_0 = \begin{bmatrix} 0 & 1 \\ -1 & 0 \end{bmatrix}$, $C_0 = \begin{bmatrix} 1 \\ 0 \end{bmatrix}^T$ are given, and $\varpi_{i1} = h_i/(\iota_1 M_i),\ \varpi_{i2} = (M_i + m_i)g/(\iota_i M_i),\ \varpi_{i3} = -h_i/M_i,\ b_i = 1/(\iota_i M_i),\ p_{i1} = (\alpha_{i1} + \alpha_{i2})/M_i,\ p_{i2} = \alpha_{i2}/(\iota_i M_i)$.

Note that the initial states of the variables designed in Theorems 4.1 and 4.2 are all taken at random and the relevant parameters are chosen as $o_{ij} = 1,\ \kappa_{ij} = 0.01,\ \mu_{ij} = 0.01,\ \beta_{ij} = 1,\ \gamma_{ij} = 1,\ \iota = 1,\ \iota_1 = 4.5,\ \bar{\beta}_i = 3,\ \bar{\gamma}_i = 0.01$,

$$P_1 = \begin{bmatrix} 26.6755 & 17.2988 & 31.7997 & 3.3063 \\ 17.2988 & 17.6746 & 46.7954 & 4.6437 \\ 31.7997 & 46.7954 & 248.2027 & 28.9702 \\ 3.3063 & 4.6437 & 28.9702 & 6.1247 \end{bmatrix},$$

$$P_2 = \begin{bmatrix} 27.5802 & 18.5661 & 36.7335 & 4.0070 \\ 18.5661 & 19.6080 & 54.8203 & 5.7964 \\ 36.7335 & 54.8203 & 298.3138 & 38.6423 \\ 4.0070 & 5.7964 & 38.6423 & 8.2636 \end{bmatrix},$$

$$P_3 = \begin{bmatrix} 28.0270 & 19.2061 & 39.2687 & 4.3647 \\ 19.2061 & 20.6122 & 59.0636 & 6.4036 \\ 39.2687 & 59.0636 & 325.7976 & 43.9995 \\ 4.3647 & 6.4036 & 43.9995 & 9.4603 \end{bmatrix},$$

$$P_4 = \begin{bmatrix} 28.4697 & 19.8495 & 41.8472 & 4.7277 \\ 19.8495 & 21.6399 & 63.4575 & 7.0320 \\ 41.8472 & 63.4575 & 354.9100 & 49.7114 \\ 4.7277 & 7.0320 & 49.7114 & 10.7441 \end{bmatrix},$$

and $\bar{J}_i,\ i \in I[1, 4]$ are selected by pole assignment so that the poles are $-1,\ -2,\ -3,\ -4$, respectively.

The DoS attacks for the edges $(1, 3)((3, 1)),\ (2, 4)((4, 2))$, and $(3, 4)((4, 3))$ are $[0.1, 1.5) \cup [3, 5.5) \cup [6.5, 8.5),\ [1.8, 2) \cup [8, 10) \cup [21, 25),\ [2, 4) \cup [35, 38) \cup [40, 45)$, and assume the edge $(0, 1)$ is secure. The agents 3 and 4 are fault free and the actuator faults of 1th and 2th agents are set as $\Lambda_1 = 0.02$ and $\Lambda_2 = 0.12$ for $50s \le t \le 70s$. Executing the controller designed in this chapter, the following simulation results are obtained.

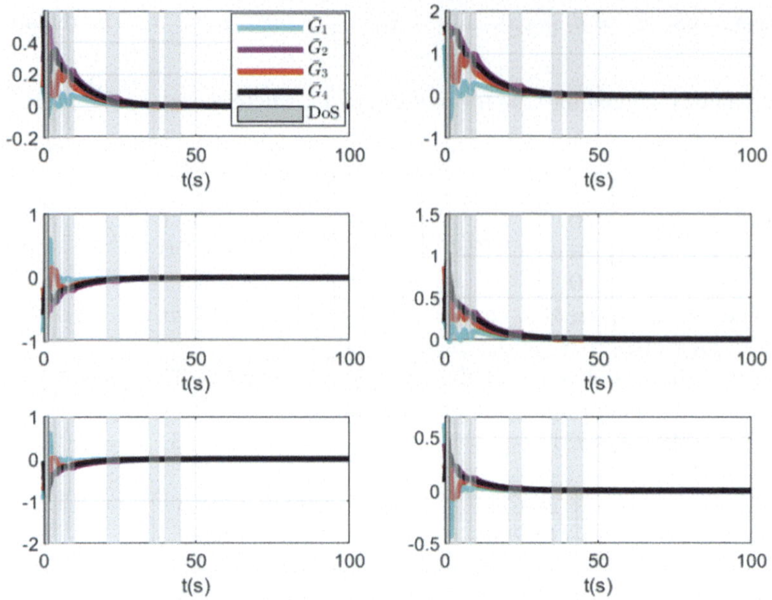

Fig. 4.1 The dynamic response of the observer error $\bar{G}_i$, reprinted from ref. Zhang et al. (2023b), copyright 2024, with permission from IEEE

Figure 4.1 depicts the dynamic response of the dynamic response of the observer error $\bar{G}_i$, $i = 1, \ldots, 4$, in (4.7a), where the shaded area indicates the attack interval. It follows from Fig. 4.1 that the proposed observer can successfully estimate the leader's system matrix. Figure 4.2 depicts the dynamic response of observer error $\bar{\eta}_i$, $i = 1, \ldots, 4$, in (4.7b), which shows that the observer error $\bar{\eta}_i$ eventually converges to 0, i.e., the observer can successfully estimate the state of the leader. Figure 4.3 depicts the dynamic response of the adaptive weight c_{ij}. Obviously, the adaptive weight eventually stabilizes and converges to the finite value. Figure 4.4 depicts the dynamic response of the output regulation error $\bar{y}_i$. As can be seen from Fig. 4.4, the multiagent system has been stabilized in 50s. At 50s, actuator faults occur in agents 1 and 2, and after a period of time, the output regulation error finally converges to zero. Thus, the cooperative output regulation problem is realized under the proposed method. Furthermore, we give a table to count the number of triggers of events under the proposed triggering mechanisms in Eqs. (4.11) and (4.36). From the statistics of the number of triggering times in Table 4.1, we can also see that the agents achieve asynchronous event-triggered communication among their neighbors. Therefore, compared with the synchronous communication in Qian et al. (2021), the event-triggered mechanisms proposed in this paper can further save the communication resources.

The multiagent system considered in Example 4.1 consists of four followers with the same number of dimensions, and to demonstrate the generality of the methods

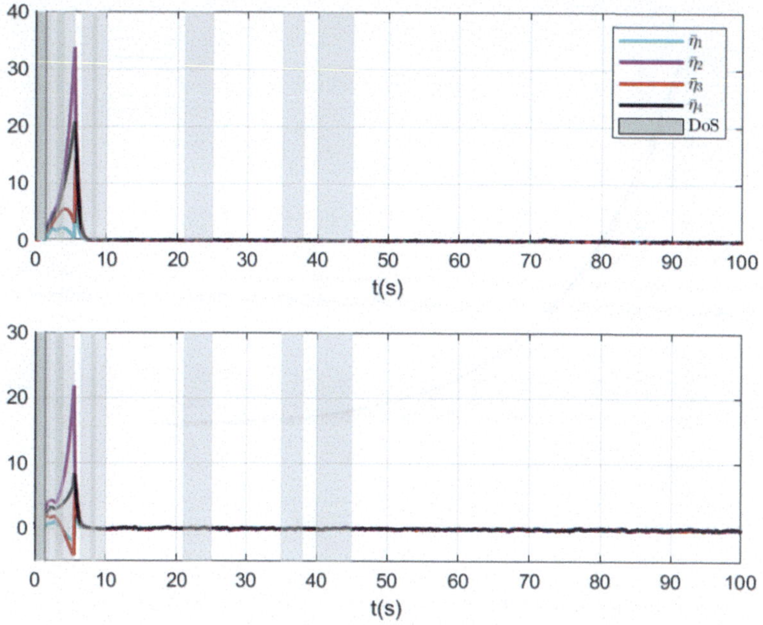

Fig. 4.2 The dynamic response of the observer error $\bar{\eta}_i$, reprinted from ref. Zhang et al. (2023b), copyright 2024, with permission from IEEE

in this chapter, multiagent systems with followers of different dimensions are considered in Example 4.2.

Example 4.2 Consider a heterogeneous multiagent system composed of five different dimensional subsystems, where the system dynamics is modeled by

$$A_i = \begin{bmatrix} 0 & 1 & 0 \\ 0 & 0 & c_i \\ 0 & -d_i & -a_i \end{bmatrix},\ B_i = \begin{bmatrix} 0 \\ 0 \\ b_i \end{bmatrix},\ E_i = \begin{bmatrix} -0.5i & 0 \\ -1 & 0.5i \\ 0 & 0 \end{bmatrix},\ C_i = \begin{bmatrix} 1 & 0 & 0 \end{bmatrix},\ i = 1, 2,$$

$$A_i = \begin{bmatrix} 0 & 1 \\ d_i & c_i \end{bmatrix}, B_i = \begin{bmatrix} 0 \\ b_i \end{bmatrix},\ E_i = \begin{bmatrix} 0.5i & 0 \\ 0 & 0 \end{bmatrix},\ C_i = \begin{bmatrix} 1 & 0 \end{bmatrix}, i = 3, 4.$$

The relative parameters $\{a_i, b_i, c_i, d_i\}$ are selected from the set $\{1, 8, 1, 6\}$, $\{1, 6, 2, 1\}$, $\{2, 5, 1, 8\}$, $\{1, 5, 1, 8\}$, and $A_0 = \begin{bmatrix} 0 & 1 \\ -1 & 0 \end{bmatrix}$, $C_0 = \begin{bmatrix} 1 \\ 0 \end{bmatrix}^T$. The leader-follower matrix of the communication topology, the DoS attack intervals, actuator faults, and other relative parameters are the same as Example 4.1.

Executing the observer and controller proposed in this chapter, the simulation results are obtained as shown in Fig. 4.5. Figure 4.5 depicts the dynamic response

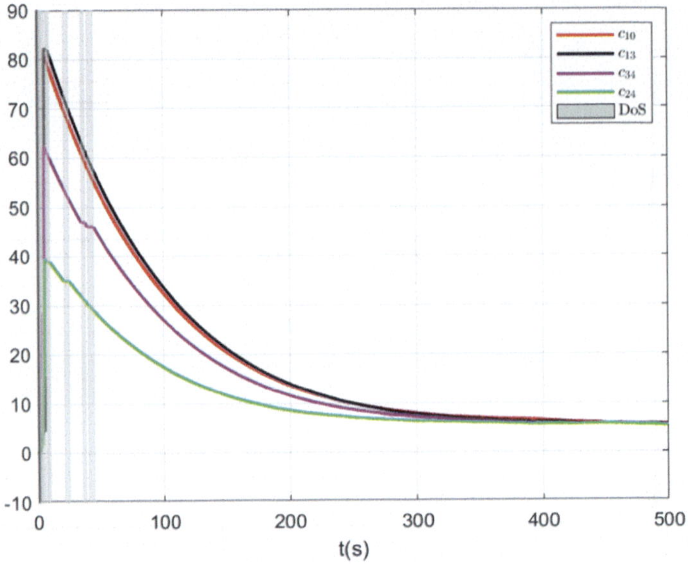

Fig. 4.3 The dynamic response of the adaptive weight c_{ij}, reprinted from ref. Zhang et al. (2023b), copyright 2024, with permission from IEEE

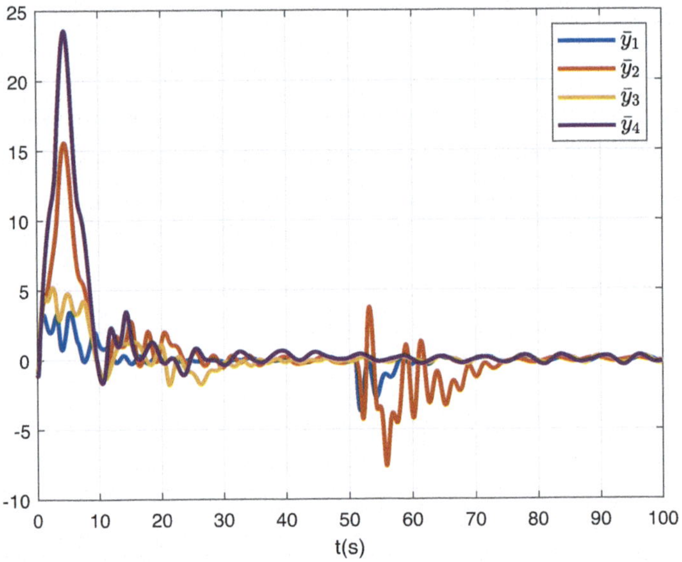

Fig. 4.4 The dynamic response of the regulated output error $\bar{y}_i$, $i = 1, 2, 3, 4$, reprinted from ref. Zhang et al. (2023b), copyright 2024, with permission from IEEE

of the output regulation error $\bar{y}_i$, which shows that the output regulation problem of multiagent system under DoS attacks and actuator faults is addressed.

Table 4.1 The number of triggers within 100 s (CTT represents the number of triggers at the communication and controller ends, sampling step size $h = 0.001$ s), reprinted from ref. Zhang et al. (2023b), copyright 2024, with permission from IEEE

Agent	1	2	3	4			
CTT in Qian et al. (2021)	325	415	492	443			
UTT (4.36)	1314	936	781	1102			
Edge	(1,0)	(1,3)	(3,1)	(2,4)	(4,2)	(3,4)	(4,3)
CTT (4.11)	303	369	398	279	222	380	362

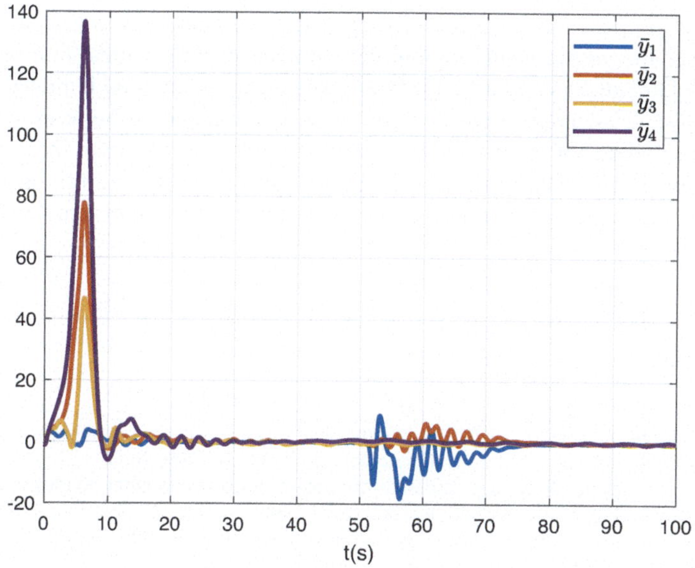

Fig. 4.5 The dynamic response of the output regulation error $\bar{y}_i$, $i = 1, 2, 3, 4$, reprinted from ref. Zhang et al. (2023b), copyright 2024, with permission from IEEE

Taking two examples together, the control algorithm proposed in this chapter can solve the output regulation problem and realize the intermittent asynchronous communication of the neighbors and intermittent update of the controller, i.e., reduce the frequency of the communication from the sensor to the controller and from the controller to the actuator. In addition, the method in this chapter not only ensures the safe operation of the network and maximizes the loss recovery in case of hacking, but also stabilizes the system in case of actuator fault. Although the considered problem in Deng and Wen (2020) and Deng et al. (2022) is the same as the one considered in this chapter, the proposed method not only relies on the global information of the communication topology graph, but also requires continuous communication between the neighbors and continuous controller updates. Therefore, the approach in this chapter is more economical in terms of system communication and computational resources compared to the approach in Deng and Wen (2020) and Deng et al. (2022). In addition, the design of both the observer and the controller does not depend on the global information of the communication topology graph. Therefore, the method in this chapter is scalable and adaptable to heterogeneous linear multiagent systems with large-scale networks.

4.5 Summary

This chapter proposed a new adaptive event-triggered fault-tolerant control scheme, which solved the output regulation problem of heterogeneous linear multiagent systems under DoS attacks and actuator faults. The proposed method included multiple advantages such as intermittent asynchronous communication between neighbors, intermittent updates of the controllers without chattering behavior, independent of global information of communication topology, and scalability. In addition, the effectiveness of the proposed method was well demonstrated through simulation examples.

References

Cai H, Lewis F, Hu G, Huang J (2017) The adaptive distributed observer approach to the cooperative output regulation of linear multi-agent systems. Automatica 75:299–305

Chen B, Ho DWC, Zhang W-A, Yu L (2019) Distributed dimensionality reduction fusion estimation for cyber-physical systems under dos attacks. IEEE Trans Syst Man Cybern Syst 49(2):455–468

Cheng B, Li Z (2019) Coordinated tracking control with asynchronous edge-based event-triggered communications. IEEE Trans Autom Control 64(10):4321–4328

Deng C, Wen C (2020) Distributed resilient observer-based fault-tolerant control for heterogeneous multiagent systems under actuator faults and dos attacks. IEEE Trans Control Netw Syst 7(3):1308–1318

Deng C, Jin X-Z, Che W-W, Wang H (2021) Learning-based distributed resilient fault-tolerant control method for heterogeneous mass under unknown leader dynamic. IEEE Trans Neural Netw Learn Syst 1–10

Deng C, Zhang D, Feng G (2022) Resilient practical cooperative output regulation for mass with unknown switching exosystem dynamics under dos attacks. Automatica 139:110172

Feng Z, Hu G (2020) Secure cooperative event-triggered control of linear multiagent systems under dos attacks. IEEE Trans Control Syst Technol 28(3):741–752

He W, Mo Z, Han Q-L, Qian F (2020) Secure impulsive synchronization in Lipschitz-type multi-agent systems subject to deception attacks. IEEE/CAA J Automatica Sinica 7(5):1326–1334

Jiang D, Wen G, Peng Z, Huang T, Rahmani A (2022) Fully distributed dual-terminal event-triggered bipartite output containment control of heterogeneous systems under actuator faults. IEEE Trans Syst Man Cybern Syst 52(9):5518–5531

Li X, Yang G (2012) Robust adaptive fault-tolerant control for uncertain linear systems with actuator failures. IET Control Theory Appl 6(10):1544–1551

Liu C, Jiang B, Wang X, Yang H, Xie S (2022) Distributed fault-tolerant consensus tracking of multi-agent systems under cyber-attacks. IEEE/CAA J Automatica Sinica 1–12

Li W, Zhang H, Zhou Y, Wang Y (2022) Bipartite formation tracking for multi-agent systems using fully distributed dynamic edge-event-triggered protocol. IEEE/CAA J Automatica Sinica 1–7

Qian Y-Y, Liu L, Feng G (2021) Cooperative output regulation of linear multiagent systems: an event-triggered adaptive distributed observer approach. IEEE Trans Autom Control 66(2):833–840

Teixeira A, Shames I, Sandberg H, Johansson KH (2015) A secure control framework for resource-limited adversaries. Automatica 51:135–148

Xu Y, Fang M, Wu Z-G, Pan Y-J, Chadli M, Huang T (2020) Input-based event-triggering consensus of multiagent systems under denial-of-service attacks. IEEE Trans Syst Man Cybern Syst 50(4):1455–1464

Yang Y, Xu H, Yue D (2019) Observer-based distributed secure consensus control of a class of linear multi-agent systems subject to random attacks. IEEE Trans Circuits Syst I: Regular Papers 66(8):3089–3099

Yang Y, Li Y, Yue D, Tian Y-C, Ding X (2021) Distributed secure consensus control with event-triggering for multiagent systems under dos attacks. IEEE Trans Cybern 51(6):2916–2928

Ye D, Chen M-M, Yang H-J (2019) Distributed adaptive event-triggered fault-tolerant consensus of multiagent systems with general linear dynamics. IEEE Trans Cybern 49(3):757–767

Zhang J, Zhang H, Liang Y, Song W (2023a) Adaptive bipartite output tracking consensus in switching networks of heterogeneous linear multiagent systems based on edge events. IEEE Trans Neural Netw Learn Syst 34(1):79–89

Zhang J, Yang D, Zhang H, Su H (2023b) Adaptive secure practical fault-tolerant output regulation of multiagent systems with DoS attacks by asynchronous communications. IEEE Trans Netw Sci Eng 10(6):4046–4055

Zhang J, Zhang H, Li W, Li K (2021) Distributed edge-event triggered consensus control for multi-agent systems by edge-based asynchronous communications. Appl Math Comput 397:125920

Zhao X, Zou S, Ma Z (2021) Decentralized resilient h_∞ load frequency control for cyber-physical power systems under dos attacks. IEEE/CAA J Automatica Sinica 8(11):1737–1751

Chapter 5
Containment Control of Multiagent Systems Under Directed Topology

5.1 Introduction

The cooperative control problems of multiagent systems considered in the previous chapters contain only one leader or no leader. However, some specific application scenarios may require multiple leaders to collaborate in order to accomplish the given task. For example, in hazardous materials handling, forest fire rescue, and so on, containment control provides an efficient solution strategy in Cao et al. (2011), Ji et al. (2008). Containment control denotes that drives all the followers to converge into a convex hull formed by all the leaders in the framework of multiagent systems. Although the research on containment control for multigent systems has achieved fruitful results, it is not easy to find out how to construct the control protocol to achieve the containment problem of multiagent systems. The authors in Li et al. (2013b) design a distributed continuous-time control algorithm and study the containment control problem of linear multiagent systems under directed graphs, where the designed control protocol relies on the eigenvalue information of the Laplace matrix (global information). In order to avoid using the global information, the authors in Li et al. (2015) design a distributed adaptive continuous-time control protocol to solve the containment control problem of multiagent systems with bounded leader inputs. It is important to emphasize that non-zero leader inputs can adjust their trajectories to avoid obstacles thus creating the desirable safety zone. In addition, since non-zero inputs from the leaders must be handled in cooperative control, this can add an extra level of difficulty, which makes the study of the containment control problem even more challenging.

In addition, in the first three chapters of this book, the cooperative control problem of multiagent systems is considered under the assumption of undirected graph. In fact, some communication networks may be modeled as directed networks, and the control protocol under directed graph is more robust than that under undirected communication topology graph. This is because in the case of undirected graph,

J. Zhang et al., *Adaptive Collaborative Control of Multiagent Systems by Event-Triggered Mechanisms*, Intelligent Control and Learning Systems 14,
https://doi.org/10.1007/978-981-97-7866-9_5

if the external environment causes the communication network to change, such as the sending device or receiving device cannot work normally, the undirected graph becomes a directed graph, and the control protocol under undirected graph may not be applicable at this time. Because undirected graph is a special case of directed graph, it is more meaningful to consider the cooperative control problem of multiagent system under directed graph. In the case of directed topology, the asymmetry of the Laplace matrix increases the difficulty of designing adaptive event-triggered control protocol. As far as the authors know, the distributed adaptive event-triggered control protocol designed to solve the cooperative control problem of multiagent systems under directed graph is still open and needs to be broken through. Note that the control protocols designed in Chaps. 2 and 3 need to be updated continuously, which may shorten the service life of agents. Although the control protocol designed in Chap. 4 meets the requirements of intermittent controller updating, it may require higher hardware requirements of system equipment, because it needs to equip event-trigger mechanism detectors at the communication end and the controller end.

Based on the above discussion, this chapter studies the containment control problem of linear multiagent systems under directed topology by adaptive event-triggered control technology. Two cases of zero input and non-zero input leaders are considered, respectively. The method in this chapter does not need to be equipped with event-triggered mechanism detectors at the communication side and the controller side.

5.2 Problem Description

Consider the multiagent system composed of M followers and N leaders. The system dynamics of ith agent can be modeled by

$$\dot{x}_i = Ax_i + Bu_i, \ i \in I[1, M+N] \tag{5.1}$$

where $x_i \in R^n$, $u_i \in R^p$ denote the state and control input for agent i, $A \in R^{n\times n}$, $B \in R^{n\times p}$ are given system matrices. Note that the follower set is labeled by $F = \{1, \cdots, M\}$ and the leader set is labeled by $L = \{M+1, \cdots, M+N\}$, and $M, \ N$ are the positive constants.

The objectives considered in this chapter are as follows.

Containment problem The containment control problem of multiagent system in Eq. (5.1) is solved under the designed controller for any initial states $x_i(0)$ $i \in I[1, M+N]$ if the followers' states asymptotically converge to the convex full formed by the leaders.

In order to realize the containment problem for multiagent system in Eq. (5.1), the following assumptions are given.

Assumption 5.1 For a directed graph $\tilde{\mathcal{G}}$, there exists a directed path from at least one leader to each follower.

Assumption 5.2 The pair (A, B) is stabilizable.

5.3 Main Results

5.3.1 The Case of the Zero Input Leaders

In this section, it is assumed that all leaders' inputs are zero. Define the relative measurement state

$$\xi_i = \sum_{j\in I[1,M]} a_{ij}\left(x_i - x_j\right) + \sum_{k\in I[M+1,M+N]} a_{ik}(x_i - x_k),\ i \in I[1, M].$$

Then, design the following adaptive event-triggered controller for follower i:

$$\begin{aligned} u_i &= \left(c_i(t_k^i) + \varphi_i(t_k^i)\right)\Xi\xi_i(t_k^i) \\ \dot{c}_i &= \xi_i^T(t_k^i)\Pi\xi_i(t_k^i) \end{aligned} \tag{5.2}$$

where c_i $(i \in I[1, M])$ denote the adaptive weights with $c_i(0) > \sqrt{2\varsigma_1}$, ς_1 is given in Eq. (5.4), φ_i $(i \in I[1, M])$ denote the nonlinear functions given by $\varphi_i = \xi_i^T P\xi_i$, Ξ, Π, and P are the feedback matrices to be designed, and the triggering sequences $\{t_k^i,\ k \in \mathbb{N},\ i \in I[1, M]\}$ $(t_0^i = 0)$ depend on the following triggering mechanism:

$$t_{k+1}^i = \inf\{t > t_k^i | h_{1i} \geq 0 \vee h_{2i} \geq 0\} \tag{5.3}$$

h_{1i}, h_{2i} are triggering functions to be designed.

Before giving the design of the triggering function, define the following measurement errors $e_i = \xi_i(t_k^i) - \xi_i$, $e_{\varphi_i} = \varphi_i(t_k^i) - \varphi_i$, $e_{c_i} = c_i(t_k^i) - c_i$, $t \in [t_k^i, t_{k+1}^i)$.

Then, design the triggering function for follower i,

$$h_{1i} = (c_i + \varphi_i)^2 e_i^T \Pi e_i - \varsigma_1\xi_i^T\Pi\xi_i - \theta_1 e^{-\iota_1 t} \tag{5.4a}$$

$$h_{2i} = (e_{c_i} + e_{\varphi_i})^2\bar{\xi}_i^T\Pi\bar{\xi}_i - \varsigma_2\xi_i^T\Pi\xi_i - \theta_2 e^{-\iota_2 t} \tag{5.4b}$$

where ς_j, θ_j, ι_j $(j = 1, 2)$ denote the positive constants and $\bar{\xi}_i = \xi_i(t_k^i)$ is given.

Remark 5.1 When executing the triggering mechanism in Eq. (5.3), it is necessary to measure continuously ξ_i. To circumvent this shortcoming, we give a new method to compute ξ_i, namely

$$\begin{aligned}\dot{\xi}_i = A\xi_i + &\sum_{j\in I[1,M]} a_{ij}B\Big(\big(c_i(t_k^i)+\varphi_i(t_k^i)\big)\Xi\xi_i(t_k^i)\\ &- a_{ij}\big(c_j(t_{k'}^j)+\varphi_j(t_{k'}^j)\big)\Xi\xi_j(t_{k'}^j)\Big)\\ &+ \sum_{k\in I[M+1,M+N]} a_{ik}B\big(c_i(t_k^i)+\varphi_i(t_k^i)\big)\Xi\xi_i(t_k^i),\end{aligned}$$

from which we get the following formula when $t \in [t_k^i, t_{k+1}^i)$,

$$\begin{aligned}\xi_i(t) = e^{A(t-t_k^i)}\xi_i(t_k^i) + &\sum_{j\in I[1,M]} a_{ij}\int_{t_k^i}^{t} e^{A(t-\tau)}B\Big(\big(c_i(t_k^i)\\ &+\varphi_i(t_k^i)\big)\Xi\xi_i(t_k^i) - \big(c_j(t_{k'}^j)+\varphi_j(t_{k'}^j)\big)\Xi\xi_j(t_{k'}^j)\Big)d\tau\\ &+ \sum_{k\in I[M+1,M+N]} a_{ik}\int_{t_k^i}^{t} e^{A(t-\tau)}B\big(c_i(t_k^i)\\ &+\varphi_i(t_k^i)\big)\Xi\xi_i(t_k^i)d\tau,\end{aligned} \tag{5.5}$$

where $t_{k'}^j$ denotes the latest triggering instant for follower j in the interval $[t_k^i, t_{k+1}^i)$. Furthermore, one can calculate $c_j(t)$ as $c_j(t) = c_j(t_k^i) + \xi_i^T(t_k^i)\Xi\xi_i(t_k^i)(t - t_k^i)$, for $t \in [t_k^i, t_{k+1}^i)$.

Inspired by Cheng and Ugrinovskii (2016), we give the following steps to compute the measurement state ξ_i, $i \in I[1, M]$.

Step 1: (a) Let $t_k^i = 0,\ k = 0,\ e_i(0) = 0,\ e_{\varphi i}(0) = 0,\ e_{ci}(0) = 0$. (b) Measure initial measurement state $\xi_i(0)$ and send $\xi_i(0)$ to its neighbor agent l, $i \in \mathcal{N}_l$.

Step 2: If the triggering conditions in the triggering mechanism in Eq. (5.3) are not met, calculate $\xi_i(t)$ by (5.5) with the latest received $\xi_j(t_{k'}^j),\ j \in \mathcal{N}_i$, and update $e_i(t),\ e_{\varphi_i}(t),\ e_{c_i}(t)$.

Step 3: If the triggering conditions in the triggering mechanism in Eq. (5.3) are met, update $k = k + 1$, $t_k^i = t,\ e_i(t) = 0,\ e_{\varphi_i}(t) = 0,\ e_{c_i}(t) = 0$, measure $\xi_i(t)$, set $\xi_i(t_k^i) = \xi_i(t)$, and send $\xi_i(t_k^i)$ to the agent l, $i \in \mathcal{N}_l$. Therefore, according to Eq. (5.5), it is known that each follower only needs to measure the relative information between the neighbors at the respective triggering moment, avoiding the need to measure the relative information continuously.

For better understanding, Fig. 5.1 depicts the structure of the event-triggered control scheme. Note that the approach in this chapter requires each agent to be equipped with sensors to measure relative information. For example, in a row of vehicles, measurement sensors need to be embedded in each vehicle to obtain the relative position and speed between the vehicles.

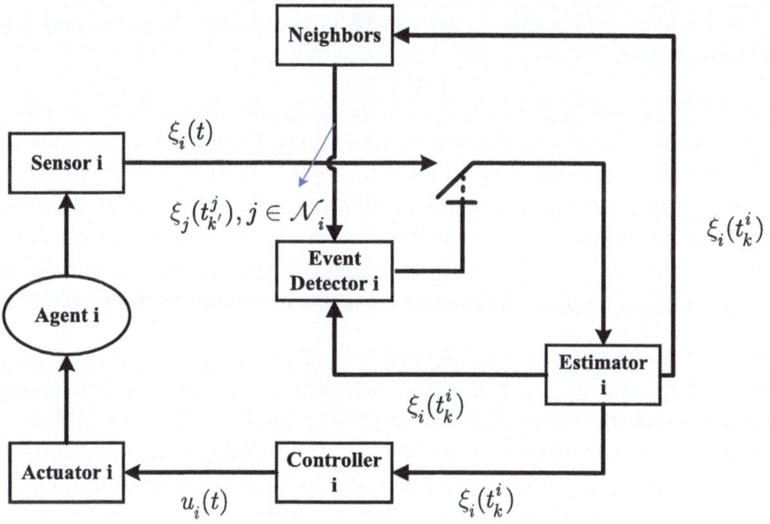

Fig. 5.1 The event-triggered control scheme for agent i, $i \in I[1, M]$, reprinted from Zhang et al. (2023), copyright 2024, with permission from IEEE

It follows from Eqs. (5.1) and (5.2) that the dynamics of the closed-loop system

$$
\begin{aligned}
\dot{x}_i &= Ax_i + \left(c_i(t^i_k) + \varphi_i(t^i_k)\right) B\Xi\xi_i(t^i_k) \\
\dot{c}_i &= \xi_i^T(t^i_k)\Pi\xi_i(t^i_k),\ i \in I[1, M] \\
\dot{x}_k &= Ax_k,\ k \in I[M+1, M+N].
\end{aligned}
$$

For the sake of writing simplicity, the definition of the following variables is given. $x_F = \mathrm{col}(x_1, \cdots, x_M)$, $x_L = \mathrm{col}(x_{M+1}, \cdots, x_N)$, $\xi = \mathrm{col}(\eta_1, \cdots, \eta_M)$, $\bar{\xi} = \mathrm{col}(\bar{\xi}_1, \cdots, \bar{\xi}_M)$ with $\bar{\xi}_i = \xi_i(t^i_k)$, $\bar{\varphi} = \mathrm{diag}\{\bar{\varphi}_1, \cdots, \bar{\varphi}_M\}$ with $\bar{\varphi}_i = \varphi_i(t^i_k)$, $\bar{c} = \mathrm{diag}\{\bar{c}_1, \cdots, \bar{c}_M\}$ with $\bar{c}_i = c_i(t^i_k)$, $\varphi = \mathrm{diag}\{\varphi_1, \cdots, \varphi_M\}$, $c = \mathrm{diag}(c_1, \cdots, c_M)$, $e = \mathrm{col}(e_1, \cdots, e_M)$, $e_c = \mathrm{diag}\{e_{c_1}, \cdots, e_{c_M}\}$, $e_\varphi = \mathrm{diag}\{e_{\varphi_1}, \cdots, e_{\varphi_M}\}$.

Based on the above definition, one can obtain that

$$
\begin{aligned}
\dot{x}_F &= (I_M \otimes A)x_F + \left((\bar{c} + \bar{\varphi}) \otimes B\Xi\right)\bar{\xi} \\
\dot{x}_L &= (I_N \otimes A)x_L \\
\xi &= (\mathcal{L}_F \otimes I_n)x_F + (\mathcal{L}_L \otimes I_n)x_L.
\end{aligned}
\tag{5.6}
$$

According to Eq. (5.6), we can calculate the derivative of ξ

$$
\dot{\xi} = (I \otimes A)\xi + \left(\mathcal{L}_F(\bar{c} + \bar{\varphi}) \otimes B\Xi\right)\bar{\xi}. \tag{5.7}
$$

It follows from (5.6) that $x_F \to -(\mathcal{L}_F^{-1}\mathcal{L}_L \otimes I)x_L$ as $t \to \infty$ if $\xi \to 0$ as $t \to \infty$. Therefore, we can conclude that when $x_F \to -(\mathcal{L}_F^{-1}\mathcal{L}_L \otimes I)x_L$, the containment control problem of multiagent system in (5.1) is solved. Thus, we call ξ as the containment error.

Remark 5.2 The nonlinear function φ_i in the controller design provides more flexibility to solve the containment problem. Obviously, the nonlinear function φ_i converges to zero as the containment error ξ converges to zero. It is emphasized that the controller designed in this chapter is inspired from Li et al. (2021), with the difference that our controller in Eq. (5.2) is based on the event-triggered mechanism and avoids the use of continuous state information. Therefore, the proposed controller avoids frequent updates and is more economical in terms of communication resources.

Remark 5.3 The proposed triggering mechanism in Eq. (5.3) consists of composite triggering conditions, i.e., $h_{1i} \geq 0$ or $h_{2i} \geq 0$. When $h_{1i} \geq 0$ or $h_{2i} \geq 0$, the triggering instant is updated, otherwise it remains unchanged. The authors in Zhang et al. (2020), Cheng and Li (2019b), Yang et al. (2018), Zhang et al. (2021) employed adaptive event-triggered control to realize containment problem of multiagent systems under undirected graph, where the single triggering condition was designed. However, for directed graph, the single triggering condition is theoretically insufficient to realize the containment control problem, which indicates that adaptive event-triggered control under directed graphs is more complex than in the case of undirected graphs.

Remark 5.4 It is worth emphasizing that all measurement errors e_i, e_{c_i}, e_{φ_i} are reset to zero at the triggering instants. Moreover, at the triggering instant t_k^i, the following formulas hold:

$$0 = \left(\bar{c}_i + \bar{\varphi}_i\right)^2 e_i(t_k^i)^T \Pi e_i(t_k^i) < \varsigma_1 \xi_i(t_k^i)^T \Pi \xi_i(t_k^i) + \theta_1 e^{-\iota_1 t_k^i}$$

or

$$0 = \left(e_{c_i}(t_k^i) + e_{\varphi_i}(t_k^i)\right)^2 \bar{\xi}_i^T \Pi \bar{\xi}_i < \varsigma_2 \xi_i^T(t_k^i) \xi_i(t_k^i) + \theta_2 e^{-\iota_2 t_k^i}.$$

Since ξ_i and $e^{-\iota_j t}(j = 1, 2)$ are continuous with respect to time t, one can obtain that

$$0 = \left(c_i(t_k^i) + \varphi_i(t_k^i)\right)^2 e_i(t_k^i)^T \Pi e_i(t_k^i) < \varsigma_1 \xi_i(t_k^i)^T \Pi \xi_i(t_k^i) + \theta_1 e^{-\iota_1 t_k^i}$$

and

$$0 = \left(e_{c_i}(t_k^i) + e_{\varphi_i}(t_k^i)\right)^2 \bar{\xi}_i^T \Pi \bar{\xi}_i < \varsigma_2 \xi_i^T(t_k^i) \xi_i(t_k^i) + \theta_2 e^{-\iota_2 t_k^i}$$

hold for a sustained period or forever. Until either of the following formulas holds true,

$$(c_i+\varphi_i)^2 e_i^T \Pi e_i - \varsigma_1 \xi_i^T \Pi \xi_i - \theta_1 e^{-\iota_1 t} = 0$$
$$(e_{c_i}+e_{\varphi_i})^2 \bar{\xi}_i^T \Pi \bar{\xi}_i - \varsigma_2 \xi_i^T \Pi \xi_i - \theta_2 e^{-\iota_2 t} = 0$$

the agent will trigger next the event.

The main results of this section are given below.

Theorem 5.1 *Under the condition of Assumptions 5.1 and 5.2 hold, select the feedback gain matrices* Ξ, Π *as* $\Xi = -B^T P$, $\Pi = PBB^T P$, *where* P *is the positive definite matrix satisfying*

$$PA + A^T P - PBB^T P + I = 0. \tag{5.8}$$

Execute adaptive event-triggered controller in (5.2), *the containment error* ξ *converges to zero and the adaptive weights* c_i $(i \in I[1, M])$ *converge to some finite positive constants, namely, the containment problem of multiagent system in Eq.* (5.1) *is solved. In addition, the multiagent system does not exhibit the Zeno behavior under the proposed triggering mechanism in Eq.* (5.3).

Proof Consider the following Lyapunov function:

$$V = \frac{1}{2}\sum_{i\in[1,M]} h_i(2c_i+\varphi_i)\varphi_i + \frac{1}{2}\sum_{i\in[1,M]} h_i(c_i-\alpha)^2 \tag{5.9}$$

where h_i $(i \in I[1, M])$ is given in Lemma 1.3, $\alpha > 0$ is a positive constant to be given. Calculate the derivative of V

$$\begin{aligned}
\dot{V} &= \sum_{i\in I[1,M]} h_i(c_i+\varphi_i)\dot{\varphi}_i + \sum_{i\in I[1,M]} h_i\varphi_i\dot{c}_i + \sum_{i\in I[1,M]} h_i(c_i-\alpha)\dot{c}_i \\
&= \xi^T\Big((c+\varphi)H\otimes(PA+A^TP)\Big)\xi - \xi^T\Big((c+\varphi)\bar{H}(\tilde{c}+\tilde{\varphi})\otimes\Pi\Big)\bar{\xi} \\
&\quad + \bar{\xi}^T(\varphi H\otimes\Pi)\bar{\xi} + \bar{\xi}^T\Big((cH-\alpha H)\otimes\Pi\Big)\bar{\xi}
\end{aligned} \tag{5.10}$$

where $\bar{H} = H\mathcal{L}_F + \mathcal{L}_F^T H$ is given, and the establishment interval satisfies $[t_p, t_{p+1})$ with $p \in \mathbb{K}$, and the time sequence $\{t_p\}$ denotes the triggering instants for the multiagent system that triggered at least one agent. Consider the second term of Eq. (5.10)

$$\begin{aligned}
&- \xi^T\Big((c+\varphi)\bar{H}(\tilde{c}+\tilde{\varphi})\otimes\Pi\Big)\bar{\xi} \\
&= -\xi^T\Big((c+\varphi)\bar{H}(c+\varphi)\otimes\Pi\Big)\bar{\xi} - \xi^T\Big((c+\varphi)\bar{H}(e_c+e_\varphi)\otimes\Pi\Big)\bar{\xi} \\
&= -\xi^T\Big((c+\varphi)\bar{H}(c+\varphi)\otimes\Pi\Big)\xi - \xi^T\Big((c+\varphi)\bar{H}(c+\varphi)\otimes\Pi\Big)e \\
&\quad - \xi^T\Big((c+\varphi)\bar{H}(e_c+e_\varphi)\otimes\Pi\Big)\bar{\xi}
\end{aligned}$$

$$\begin{aligned}&\leq -\lambda_1(\bar{H})\xi^T\Big((c+\varphi)^2\otimes\Pi\Big)\xi-\xi^T\Big((c+\varphi)\bar{H}(c+\varphi)\otimes\Pi\Big)e\\&-\xi^T\Big((c+\varphi)\bar{H}(e_c+e_\varphi)\otimes\Pi\Big)\bar{\xi}.\end{aligned}\tag{5.11}$$

It follows from Lemma 1.6 that

$$\begin{aligned}-\xi^T\Big((c+\varphi)\bar{H}(c+\varphi)\otimes\Pi\Big)e\leq&\frac{\lambda_1(\bar{H}^2)}{2\beta_1}\xi^T\Big((c+\varphi)^2\otimes\Pi\Big)\xi\\&+\frac{\beta_1}{2}e^T\Big((c+\varphi)^2\otimes\Pi\Big)e\end{aligned}\tag{5.12}$$

and

$$\begin{aligned}-\xi^T\Big((c+\varphi)\bar{H}(e_c+e_\varphi)\otimes\Pi\Big)\bar{\xi}\leq&\frac{\lambda_1(\bar{H}^2)}{2\beta_2}\xi^T\Big((c+\varphi)^2\otimes\Pi\Big)\xi\\&+\frac{\beta_2}{2}\bar{\xi}^T\Big((e_c+e_\varphi)^2\otimes\Pi\Big)\bar{\xi}.\end{aligned}\tag{5.13}$$

Consider the last two terms of Eq. (5.10)

$$\begin{aligned}&\bar{\xi}^T(\varphi H\otimes\Pi)\bar{\xi}+\bar{\xi}^T\Big((cH-\alpha H)\otimes\Pi\Big)\bar{\xi}\\&=(e+\xi)^T(\varphi H\otimes\Pi)(e+\xi)+(e+\xi)^T\Big((cH-\varphi G)\otimes\Pi\Big)(e+\xi)\\&\leq 2e^T\Big((c+\varphi)H\otimes\Pi\Big)e+2\xi^T\Big((c+\varphi)H\otimes\Pi\Big)\xi\\&+e^T(\alpha H\otimes\Pi)e-\frac{1}{2}\xi^T(\alpha H\otimes\Pi)\xi.\end{aligned}\tag{5.14}$$

According to Eqs. (5.11)–(5.14), one can obtain the following formula:

$$\begin{aligned}\dot{V}\leq&\,\xi^T\Big((c+\varphi)H\otimes(PA+A^TP)\Big)\xi-\beta_3\xi^T\Big((c+\varphi)^2\otimes\Pi\Big)\xi\\&+\frac{\beta_1}{2}e^T\Big((c+\varphi)^2\otimes\Pi\Big)e+\frac{\beta_2}{2}\bar{\xi}^T((e_c+e_\varphi)^2\otimes\Pi)\bar{\xi}\\&+2e^T\Big((c+\varphi)H\otimes\Pi\Big)e+2\xi^T\Big((c+\varphi)H\otimes\Pi\Big)\xi\\&+e^T(\alpha H\otimes\Pi)e-\frac{1}{2}\xi^T(\alpha H\otimes\Pi)\xi\end{aligned}\tag{5.15}$$

where $\beta_3=\lambda_1(\bar{H})-\frac{\lambda_1(\bar{H}^2)}{2\beta_1}-\frac{\lambda_1(\bar{H})}{2\beta_2}>0$ can be ensured by selecting the suitable parameter β_j $(j=1,2)$.

Using the event-triggered mechanism in Eq. (5.3) the following inequality can be obtained:

$$\begin{aligned}
&\frac{\beta_1}{2}e^T\Big((c+\varphi)^2\otimes\Pi\Big)e\leq\frac{\beta_1\varsigma_1}{2}\xi^T(I\otimes\Pi)\xi+\frac{\beta_1\theta_1}{2}Me^{-\iota_1 t}\\
&\frac{\beta_2}{2}\bar{\xi}^T\Big((e_c+e_\varphi)^2\otimes\Pi\Big)\bar{\xi}\leq\frac{\beta_2\varsigma_2}{2}\xi^T(I\otimes\Pi)\xi+\frac{\beta_2\theta_2}{2}Me^{-\iota_2 t}\\
&2e^T\Big((c+\varphi)H\otimes\Pi\Big)e\leq\frac{2\varsigma_1}{c_m}\xi^T(H\otimes\Pi)\xi+\frac{2\theta_1}{c_m}e^{-\iota_1 t}\\
&e^T(\alpha H\otimes\Pi)e\leq\frac{\alpha\varsigma_1}{c_m^2}\xi^T(H\otimes\Pi)\xi+\frac{\alpha\varsigma_1}{c_m^2}Me^{-\iota_1 t}
\end{aligned}\tag{5.16}$$

where $c_m=\min_{i\in I[1,M]}c_i(0)$ is given. It follows from Eq. (5.16) that Eq. (5.15) can be written as

$$\begin{aligned}
\dot{V}\leq&\,\xi^T\Big((c+\varphi)H\otimes(PA+A^TP)\Big)\xi+\xi^T\Big(\big(-\beta_3(c+\varphi)^2+\beta_4 I\\
&+2(c+\varphi)H-\beta_5 H\big)\otimes\Pi\Big)\xi+\upsilon
\end{aligned}\tag{5.17}$$

where $\beta_4=\frac{\beta_1\varsigma_1}{2}+\frac{\beta_2\varsigma_2}{2}$, $\beta_5=\frac{1}{2}\alpha-\frac{2\varsigma_1}{c_m}-\frac{\alpha}{c_m^2}$, $\upsilon=(\frac{\beta_1\theta_1}{2}+\frac{2\theta_1}{c_m}+\frac{\alpha\varsigma_1}{c_m^2})Me^{-\iota_1 t}+\frac{\beta_2\theta_2}{2}Me^{-\iota_2 t}$ are given. Obviously, $\beta_5>0$ can be ensured by selecting α. Moreover, the following formula can be obtained:

$$-\beta_3(c+\varphi)^2+\beta_4 I+2(c+\varphi)H-\beta_5 H<-(c+\varphi)H.\tag{5.18}$$

Equation (5.18) can be equivalently written as the following formula:

$$-\beta_3(c_i+\varphi_i-\frac{3h_i}{2\beta_3})^2+\frac{9h_i^2}{4\beta_3}+\beta_4-\beta_5 h_i<0.\tag{5.19}$$

Obviously, Eq. (5.19) holds if $\beta_3>\max_{i\in I[1,M]}\{\frac{9h_i}{4\beta_3}+\frac{\beta_4}{h_i}\}$.

Set $\alpha>\frac{2c_m^2}{c_m^2-2\varsigma_1}\max_{i\in I[1,M]}\{\frac{9h_i}{4\beta_3}+\frac{\beta_4}{h_i}+\frac{2\varsigma_1}{c_m}\}$. Then, combining Eqs. (5.18) and (5.19), one obtains Eq. (5.17) can be written as

$$\dot{V}\leq\xi^T\Big((c+\varphi)H\otimes(PA+A^TP-\Pi)\Big)\xi+\upsilon=-\xi^T\Big((c+\varphi)H\otimes I\Big)\xi+\upsilon.\tag{5.20}$$

Set $\beta_6=(\frac{\beta_1\theta_1}{2}+\frac{2\theta_1}{c_m}+\frac{\alpha\varsigma_1}{c_m^2})M$, $\beta_7=\frac{\beta_2\theta_2}{2}M$, $W=V+\frac{\beta_6}{\iota_1}e^{-\iota_1 t}+\frac{\beta_7}{\iota_2}e^{-\iota_2 t}$. Calculate the derivate of W as

$$\dot{W}\leq-\xi^T\Big((c+\varphi)H\otimes I\Big)\xi\leq 0\tag{5.21}$$

from which we can be obtain that W is bounded. It is further obtained that V is bounded, so $c_i,\ \varphi_i,\ \xi_i\ (i\in I[1,M])$ are bounded. Integrating on both sides of Eq. (5.21) yields

$$\int_0^t \xi^T\Big((c+\varphi)H\otimes I\Big)\xi d\tau \le W(0)-W(t)$$

which implies that $\lim_{t\to\infty}\int_0^t \xi^T\Big((c+\varphi)H\otimes I\Big)\xi ds$ is bounded. Due to the fact that $\xi^T\Big((c+\varphi)H\otimes I\Big)\xi$ is differentiable and bounded, one can obtain that $\xi^T\Big((c+\varphi)H\otimes I\Big)\xi$ is uniformly continuous. It follows from Lemma 1.5 that $\lim_{t\to\infty}\xi^T\Big((c+\varphi)H\otimes I\Big)\xi = 0$ can be obtained. Furthermore, one can obtain that $\lim_{t\to\infty}\xi = 0$ and the adaptive weights c_i $(i\in I[1,M])$ converge to some finite constants.

Finally, we analyze that Zeno behavior does not occur in multiagent system. In this case, we adopt contradiction by assuming that the ith agent exhibits Zeno behavior. According to the definition of Zeno behavior, we can get $\lim_{k\to\infty}t_k^i = t_\infty^i < \infty$. According to the triggering mechanism in Eq. (5.3), it can be obtained that

$$\Big(c_i(t_{k+1}^i)+\varphi_i(t_{k+1}^i)\Big)^2 e_i^T(t_{k+1}^i)\Pi e_i(t_{k+1}^i)\ge\theta_1 e^{-\iota_1 t_{k+1}^i}$$

or

$$\Big(e_{c_i}(t_{k+1}^i)+e_{\varphi_i}(t_{k+1}^i)\Big)^2\bar{\xi}_i\Pi\bar{\xi}_i\ge\theta_2 e^{-\iota_2 t_{k+1}^i}.$$

According to $e_i(t_{k+1}^i)=0,\ e_{c_i}(t_{k+1}^i)=0,\ e_{\varphi_i}(t_{k+1}^i)=0$, one can obtain that

$$0<\theta_1 e^{-\iota_1 t_\infty^i}\le\Big(c_i(t_\infty^i)+\varphi_i(t_\infty^i)\Big)^2 e_i^T(t_\infty^i)\Pi e_i(t_\infty^i)=0$$

or

$$0<\theta_2 e^{-\iota_2 t_\infty^i}\le\Big(e_{c_i}(t_\infty^i)+e_{\varphi_i}(t_\infty^i)\Big)^2\bar{\xi}_i\Pi\bar{\xi}_i=0$$

which are obviously contradictory and so the assumption is wrong. Thus, the multiagent system does not exhibit the Zeno behavior for any finite time. The proof is completed. □

Remark 5.5 It is worth emphasizing that Theorem 5.1 only analyzes the fact that Zeno behavior does not occur in any finite time. In fact, many systems complete their tasks in a finite amount of time. Therefore, it makes sense to rule out Zeno behavior in finite time (Cheng and Li 2019b, a). To the best of the authors' knowledge, so far, there exist many related results that use adaptive event-triggered control schemes to study the cooperative control problem of multiagent systems, which exclude the Zeno behavior only in a finite time, such as (Li et al. 2022, 2021; Cai et al. 2022; Cheng and Li 2019a). However, if the exponential decay terms $\theta_1 e^{-\iota_1 t}$, θ_2 are replaced by the very small positive constants θ_1, θ_2, then a similar analysis can be performed to obtain that no Zeno behavior occurs in the multiagent system at any time. At this point an additional cost is required, i.e., the containment error converges to a small bounded set.

Remark 5.6 The novelties of this chapter are as follows: 1) The proposed event-triggered controller ensures both intermittent communication among the agents and intermittent updating of the controllers; 2) The considered communication topology is more general. The proposed method combines the advantages of event-triggered control and adaptive control. 3)The distributed adaptive event-triggered controller is designed to be updated only at the triggering instants. Its construction depends solely on the relative information among neighbors at triggering instants, avoiding the need for continuous measurement of relative information among neighbors.

Remark 5.7 The control algorithm designed in this chapter contains the parameters such as $\varsigma_1, \theta_1, \iota_1, \varsigma_2, \theta_2, \iota_2$. Theorem 5.1 shows that the containment control of the linear multiagent system in Eq. (5.1) can be realized by choosing appropriate parameters. In fact, different parameter choices can lead to different control performances. In general, the larger the parameters $\varsigma_1, \theta_1, \varsigma_2, \theta_2$, and the smaller the parameters ι_1, ι_2, the lower the communication frequency of the multigent system, but the slower the convergence speed. Therefore, the user should balance the communication frequency and convergence speed of the multiagent system by choosing appropriate parameters.

5.3.2 The Case of the Non-zero Input Leaders

In this section, it is assumed that all leaders have non-zero inputs and satisfy the following conditions.

Assumption 5.3 The inputs to the leaders u_i $(i \in I[M+1, M+N])$ are bounded and satisfy $\|u_i\| \le \gamma_i$, where $\gamma_i > 0$ is the positive constant.

Assume that each follower only knows the bound γ_i of the leaders' inputs and does not need to know the exact input u_i $(i \in I[1, M+N])$ of each leader.

Design the following distributed adaptive event-triggered controller for the ith follower:

$$u_i = \left(c_i(t_k^i) + \varphi_i(t_k^i)\right)\Xi\xi_i(t_k^i) + \nu g\big(\Xi\xi_i(t_k^i)\big) \tag{5.22}$$

where c_i and φ_i are given in Eq. (5.2), ν denotes the positive constant with $\nu > \max_{i\in I[M+1,N]}\{\gamma_i\}$, and $g(\cdot)$ denotes the nonlinear function. $g(\cdot)$ is given in the following form, for $\forall\, r \in R^p$,

$$g(r) = \begin{cases} \frac{r}{\|r\|}, & \text{if } \|r\| \neq 0 \\ 0, & \text{if } \|r\| = 0. \end{cases}$$

Design the triggering mechanism for agent i as

$$t_{k+1}^i = \inf\{t > t_k^i | h_{1i} \geq 0 \vee h_{2i} \geq 0 \vee h_{3i} \geq 0\} \tag{5.23}$$

where $h_{1i},\ h_{2i}$ are given in (5.23) and h_{3i} is given by

$$f_{3i} = (c_i + \varphi_i)\|\Xi e_i\| - \varsigma_3 \xi_i^T \Pi \xi_i - \theta_3 e^{-\iota_3 t} \tag{5.24}$$

and $\theta_3,\ \iota_3 > 0$ are the given positive constants.

Remark 5.8 In the designed distributed adaptive event-triggered controller in Eq. (5.22), $c_i(t_k^i),\ \varphi_i(t_k^i),\ \xi_i(t_k^i),\ (i \in I[N+1, N+M])$ are introduced to allow the controller to update only at its own triggering instants. For the case with non-zero leader inputs, the adaptive event-triggered control protocol in Eq. (5.2) and triggering mechanism in Eq. (5.3) are no longer applicable. In order to counteract the effects of non-zero leader inputs on the system, the nonlinear function $g(\cdot)$ is introduced into the design of the controller in Eq. (5.22) and the triggering function h_{3i} is introduced into the design of the triggering mechanism in Eq. (5.23). This ensures that the containment problem of the linear multiagent system in Eq. (5.1) is solved in this case.

In the next step, the definitions of some of the following variables will be given. $u_L = \mathrm{col}(u_{M+1}, \cdots, u_N),\ G(\Xi, \xi, e) = \mathrm{col}\Big(g(\Xi\xi_1(t_k^1)), \cdots, g(\Xi\xi_M(t_k^M))\Big)$. Thus, based on the adaptive event-triggered controller in Eq. (5.22), we can obtain that the closed-loop system satisfies that

$$\begin{aligned} \dot{x}_F &= (I_M \otimes A)x_F + \Big((\bar{c} + \bar{\varphi}) \otimes B\Xi\Big)\bar{\xi} + (\nu I_M \otimes B)G(\Xi, \xi, e) \\ \dot{x}_L &= (I_N \otimes A)x_L + (I_N \otimes B)u_L. \end{aligned} \tag{5.25}$$

Then, we can obtain the derivative of ξ

$$\begin{aligned} \dot{\xi} &= (I \otimes A)\xi + \Big(\mathcal{L}_F(\bar{c} + \bar{\varphi}) \otimes B\Xi\Big)\bar{\xi} + (\nu\mathcal{L}_F \otimes B)G(\Xi, \xi, e) \\ &\quad + (\mathcal{L}_L \otimes B)u_L. \end{aligned} \tag{5.26}$$

Theorem 5.2 *Under the condition of Assumptions 5.1–5.3, the feedback gain matrices $\Xi,\ \Pi$ are the same as Theorem 5.1. Execute adaptive event-triggered controller in Eq. (5.22), the containment error ξ converges to zero and the adaptive weights $c_i\ (i \in I[1, M])$ converge to some finite positive constants, namely, the containment problem of multiagent system in Eq. (5.1) is solved. In addition, the multiagent system does not exhibit the Zeno behavior under the proposed triggering mechanism in Eq. (5.23).*

Proof Consider the Lyapunov function. Select the Lyapunov function in Eq. (5.9) and compute its derivative as

$$
\begin{aligned}
\dot{V} &= \xi^T\Big((c+\varphi)H(PA+A^TP)\Big)\xi - \xi^T\Big((c+\varphi)\bar{H}(\bar{c}+\bar{\varphi})\otimes\Pi\Big)\bar{\xi} \\
&\quad - 2\xi^T\Big(\nu(c+\varphi)H\mathcal{L}_F\otimes\Xi^T\Big)G(\Xi,\xi,e) \\
&\quad - 2\xi^T\Big((c+\varphi)H\mathcal{L}_L\otimes\Xi^T\Big)u_L \\
&\quad + \bar{\xi}^T(\varphi H\otimes\Pi)\bar{\xi} + \bar{\xi}^T\Big((cH-\alpha H)\otimes\Pi\Big)\bar{\xi}.
\end{aligned} \tag{5.27}
$$

Consider the second term of Eq. (5.27)

$$
\begin{aligned}
& -2\xi^T\Big((c_i+\varphi_i)H\mathcal{L}_L\otimes\Xi^T\Big)u_L \\
&= -\sum_{i\in I[1,M]}\sum_{k\in I[M+1,M+N]} 2(c_i+\varphi_i)h_i a_{ik}\xi_i^T\Xi^T u_k \\
&\le 2\max_{k\in I[M+1,M+N]}\gamma_k \sum_{i\in I[1,M]}\sum_{k\in I[M+1,M+N]}(c_i+\varphi_i)h_i a_{ik}\|\Xi\xi_i\|.
\end{aligned} \tag{5.28}
$$

Consider the third term of Eq. (5.27)

$$
\begin{aligned}
& -2\xi^T\Big(\nu(c+\varphi)H\mathcal{L}_F\otimes\Xi^T\Big)G(\Xi,\xi,e) \\
&= -\sum_{i\in I[1,M]} 2\nu(c_i+\varphi_i)h_i l_{ii}\xi_i^T\Xi^T g\Big(\Xi(\xi_i+e_i)\Big) \\
&\quad + \sum_{i\in I[1,M]}\sum_{k=1,k\neq i}^{M} 2\nu(c_i+\varphi_i)h_i a_{ik}\xi_i^T\Xi^T g\Big(\Xi(\xi_k+e_k)\Big) \\
&\le -\sum_{i\in I[1,M]} 2\nu(c_i+\varphi_i)h_i l_{ii}(\|\Xi\xi_i\|-2\|\Xi e_i\|) \\
&\quad + \sum_{i\in I[,M]}\sum_{k=1,k\neq i}^{M} 2\nu(c_i+\varphi_i)h_i a_{ik}\|\Xi\xi_i\| \\
&= \sum_{i\in I[1,M]} 4\nu(c_i+\varphi_i)h_i l_{ii}\|\Xi e_i\| \\
&\quad - \sum_{i\in I[,M]}\sum_{k\in I[M+1,M+N]} 2\nu(c_i+\varphi_i)h_i a_{ik}\|\Xi\xi_i\| \\
&\le \sum_{i\in I[1,M]} 4\nu h_i l_{ii}(\varsigma_3\|\Xi\xi_i\|^2+\theta_3 e^{-\iota_3 t}) \\
&\quad - \sum_{i\in I[1,M]}\sum_{k\in I[M+1,M+N]} 2\nu(c_i+\varphi_i)h_i a_{ik}\|\Xi\xi_i\|.
\end{aligned} \tag{5.29}
$$

Substituting Eqs. (5.11)–(5.14) and (5.28)–(5.29) into (5.27) yields that

$$\begin{aligned}\dot{V} \leq\, & \xi^T\Big((c+\varphi)H \otimes (PA + A^T P)\Big)\xi - \beta_3 \xi^T\Big((c+\varphi)^2 \otimes \Pi\Big)\xi \\ & + \frac{\beta_1}{2} e^T\Big((c+\varphi)^2 \otimes \Pi\Big)e + \frac{\beta_2}{2}\bar{\xi}^T((e_c + e_\varphi)^2 \otimes \Pi)\bar{\xi} \\ & + 2e^T\Big((c+\varphi)H \otimes \Pi\Big)e + 2\xi^T\Big((c+\varphi)H \otimes \Pi\Big)\xi \\ & + e^T(\alpha H \otimes \Pi)e - \frac{1}{2}\xi^T(\alpha H \otimes \Pi)\xi + \beta_8 \varsigma_3 \xi^T(I \otimes \Pi)\xi \\ & + \beta_8 \theta_3 M e^{-\iota_3 t} \end{aligned} \tag{5.30}$$

where $\beta_8 = \max_{i \in I[1,M]}\{4\nu h_i l_{ii}\}$ is given. Let $\beta_9 = \beta_4 + \beta_8 \varsigma_3$. Taking advantage of Eq. (5.16), one can obtain that Eq. (5.30) satisfies

$$\begin{aligned}\dot{V} \leq\, & \xi^T\Big((c+\varphi)H \otimes (PA + A^T P)\Big)\xi \\ & + \xi^T\Big(\big(-\beta_3(c+\varphi)^2 + \beta_9 I + 2(c+\varphi)H - \beta_5 H\big) \otimes \Pi\Big)\xi + \upsilon_1 \end{aligned} \tag{5.31}$$

where $\upsilon_1 = \upsilon + \beta_8 \theta_3 M e^{-\iota_3 t}$ is given and relative parameters are the same as Theorem 5.1. Let $\alpha > \frac{2c_m^2}{c_m^2 - 2\varsigma_1}\max_{i \in I[1,M]}\{\frac{9h_i}{4\beta_3} + \frac{\beta_9}{g_i} + \frac{2\varsigma_1}{c_m}\}$. Similar to Eq. (5.20), we obtain

$$\dot{V} \leq -\xi^T\Big((c+\varphi)H \otimes I\Big)\xi + \upsilon_1.$$

The rest of the proof is similar to the proof of Theorem 5.1 and is omitted here to save space. The proof ends here. □

Remark 5.9 As described in Li et al. (2013a), non-zero inputs to the leaders cause the multiagent system in Eq. (5.1) heterogeneous. Therefore, the adaptive event-triggered controller (2.2) is no longer applicable. Obviously, if $\nu = 0$ is taken, the controller in (5.22) can be reduced to the controller in Eq. (5.22). In order to overcome the effect of the non-zero input on the consensus result, the nonlinear term $g(\cdot)$ is added to the controller in (2.22) and a extra triggering function h_{3i} is added to the triggering mechanism in (5.23). The introduction of nonlinear terms complicates the consensus analysis.

Remark 5.10 The containment problems for first- or second-order multiagent systems were reported in Liu et al. (2016), Xiong et al. (2022), Yang et al. (2019), where the controllers in Xiong et al. (2022), Yang et al. (2019) were updated at the triggering instants of themselves and neighbors, and the controllers in Liu et al. (2016) depended on all non-zero eigenvalues of $\mathcal{L}_F$. The containment problems of linear multiagent systems were reported in Wang et al. (2021), Zhang et al. (2021), where the controllers depended on global information and required to be updated continuously. In contrast to these mentioned works, the approach in this chapter not only realizes intermittent communication among the neighbors but also intermittent update of the controllers only at their own triggering moments. More importantly, the method in this chapter inherits the advantages of event-triggered control and adaptive control, and the proposed method is also independent of the global information of the communication topologies.

5.4 Simulation Examples

In this section, two examples are given to validate the effectiveness and merits of the methods proposed in this chapter.

Example 5.1 This example considers a multiagent system consisting of 6 followers (labeled $4, \cdots, 9$) and 3 leaders (labeled $1, 2, 3$), where their dynamics are shown in Eq. (5.1), and

$$A = \begin{bmatrix} 0 & 1 & 0 \\ 0 & 0 & 1 \\ 0 & 0 & 0 \end{bmatrix}, \quad B = \begin{bmatrix} 0 \\ 0 \\ 1 \end{bmatrix}.$$

The communication topology is shown in Fig. 5.2.

According to Algebraic Riccati equation in Eq. (5.8), we can calculate matrix P as

$$P = \begin{bmatrix} 2.4142 & 2.4142 & 1.0000 \\ 2.4142 & 4.8284 & 2.4142 \\ 1.0000 & 2.4142 & 2.4142 \end{bmatrix}.$$

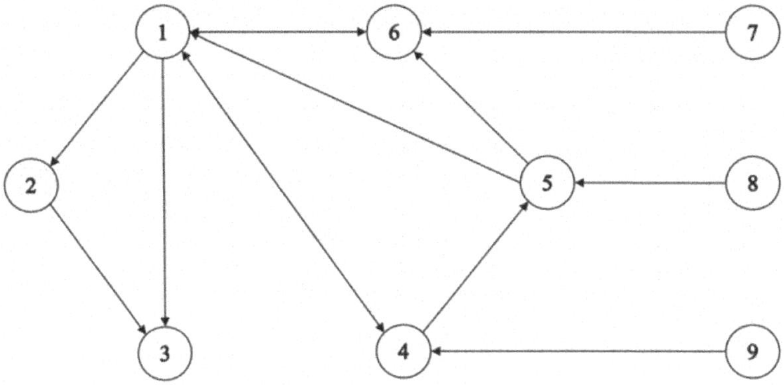

Fig. 5.2 The communication topology $\tilde{\mathcal{G}}$

Then, we can obtain that the matrices $\Xi,\ \Pi$ satisfy

$$\Xi = \begin{bmatrix} -1 \\ -2.4142 \\ -2.4142 \end{bmatrix},\ \Pi = \begin{bmatrix} 1.0000 & 2.4142 & 2.4142 \\ 2.4142 & 5.8284 & 5.8284 \\ 2.4142 & 5.8284 & 5.8284 \end{bmatrix}.$$

The initial states $x_i(0)\ i \in I[1, 9]$ and adaptive coupling weights $c_i(0)\ i \in I[1, 6]$ are randomly taken in Theorems 5.1 and 5.2. The relevant parameters in the proposed controller in Eq. (5.2) and triggering mechanism in Eq. (5.3) are taken to satisfy $\varsigma_1 = \varsigma_2 = 0.1,\ \theta_1 = \theta_2 = 1,\ \iota_1 = \iota_2 = 0.05$. The leaders' inputs are selected as $u_i = 0.1\sin(t)\ i \in I[7, 9]$, the relevant parameters in the proposed controller in Eq. (5.22) and triggering mechanism in Eq. (5.23) are taken to satisfy $\nu = 0.31,\ \varsigma_1 = \varsigma_2 = \varsigma_3 = 3,\ \theta_1 = \theta_2 = \theta_3 = 4,\ \iota_1 = \iota_2 = \iota_3 = 0.05$. Under the conditions of Theorems 5.1 and 5.2, we can obtain the simulation results as shown in Figs. 2.3–2.7.

Figure 5.3 depicts the dynamic response of the adaptive weights, from which we can see that the adaptive weights eventually converge to stable finite values under Theorems 5.1 and 5.2, respectively. Figure 5.4 depicts the dynamic response of the containment errors, from which we can see that the containment errors eventually converge to small bounded set. Thus, the containment problem of multiagent system in (5.1) under the proposed controllers and triggering mechanisms is solved. Figure 5.5 shows the triggering instants under the triggering mechanisms in Eqs. (5.3) and (5.23), from which we can obtain the multiagent system does not exhibit the Zeno behavior under Theorems 5.1 and 5.2, respectively. Figure 5.6 depicts the dynamic response of the controllers, from which we can see that the designed controllers have the ZOH properties. In addition, it is observed from Fig. 5.8 that all the third dimensions of the follower states end up in the convex packet formed by the third dimensions of the leader states. In summary, Figs. 5.3, 5.4, 5.5, 5.6, 5.7, and 5.8 show the feasibility of the proposed control algorithm.

To further demonstrate the advantages of the methods in this chapter, comparative simulation is performed.

Example 5.2 This example compares the proposed control protocol with those in Li et al. (2021). We consider the case where there is only one leader labeled 7 and 6 followers labeled $1, \cdots, 6$. The corresponding adjacency matrix of the communication topology satisfies

$$\mathcal{A} = \begin{bmatrix} 0&1&0&0&0&0&0 \\ 0&0&0&0&0&0&1 \\ 1&0&0&0&0&1&0 \\ 0&0&1&0&0&1&0 \\ 0&0&0&0&0&0&1 \\ 0&0&0&0&1&0&0 \\ 0&0&0&0&0&0&0 \end{bmatrix}.$$

Choose the same system matrices as in Example 5.1. The control input for the leader is selected as $u_i(t) = 0.1 * sin(t)$. The matrices Ξ and Π are the same as in Example

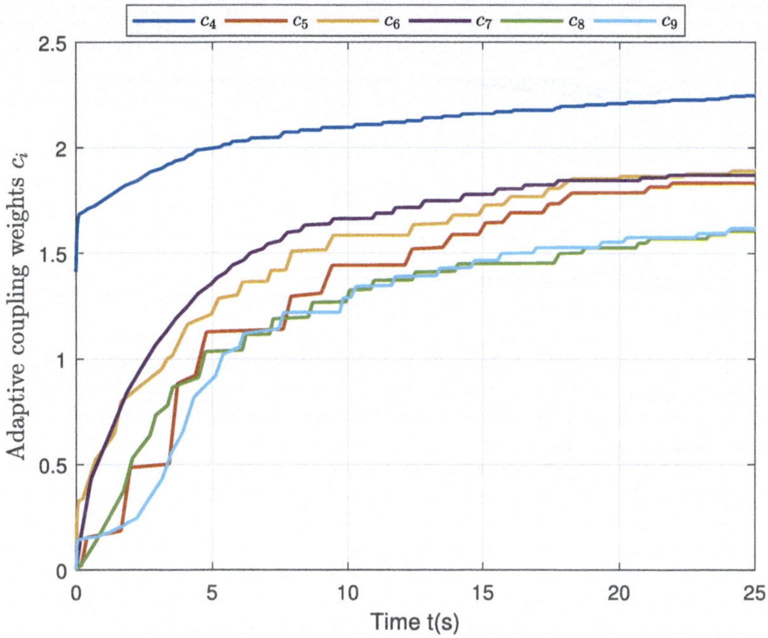

(a)

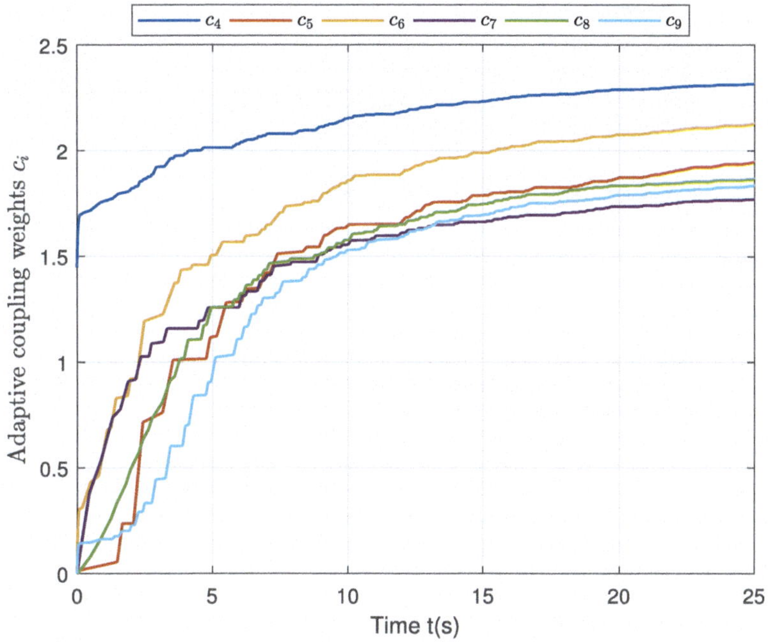

(b)

Fig. 5.3 **a** Theorem 5.1: The trajectories of the adaptive weights c_i, $i \in I[1, 6]$, **b** Theorem 5.2: The trajectories of the adaptive weights c_i, $i \in I[1, 6]$, reprinted from Zhang et al. (2023), copyright 2024, with permission from IEEE

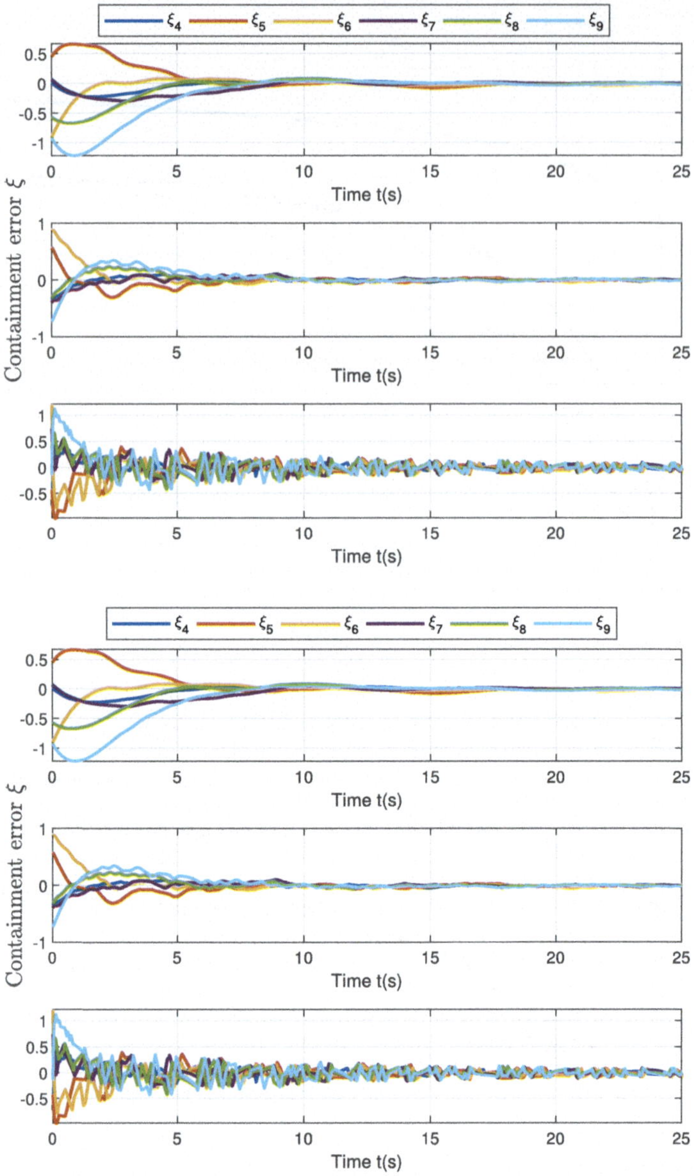

Fig. 5.4 **a** Theorem 5.1: The trajectories of the containment errors ξ_i, $i \in I[1, 6]$, **b** Theorem 5.2: The trajectories of the containment errors ξ_i, $i \in I[1, 6]$, reprinted from Zhang et al. (2023), copyright 2024, with permission from IEEE

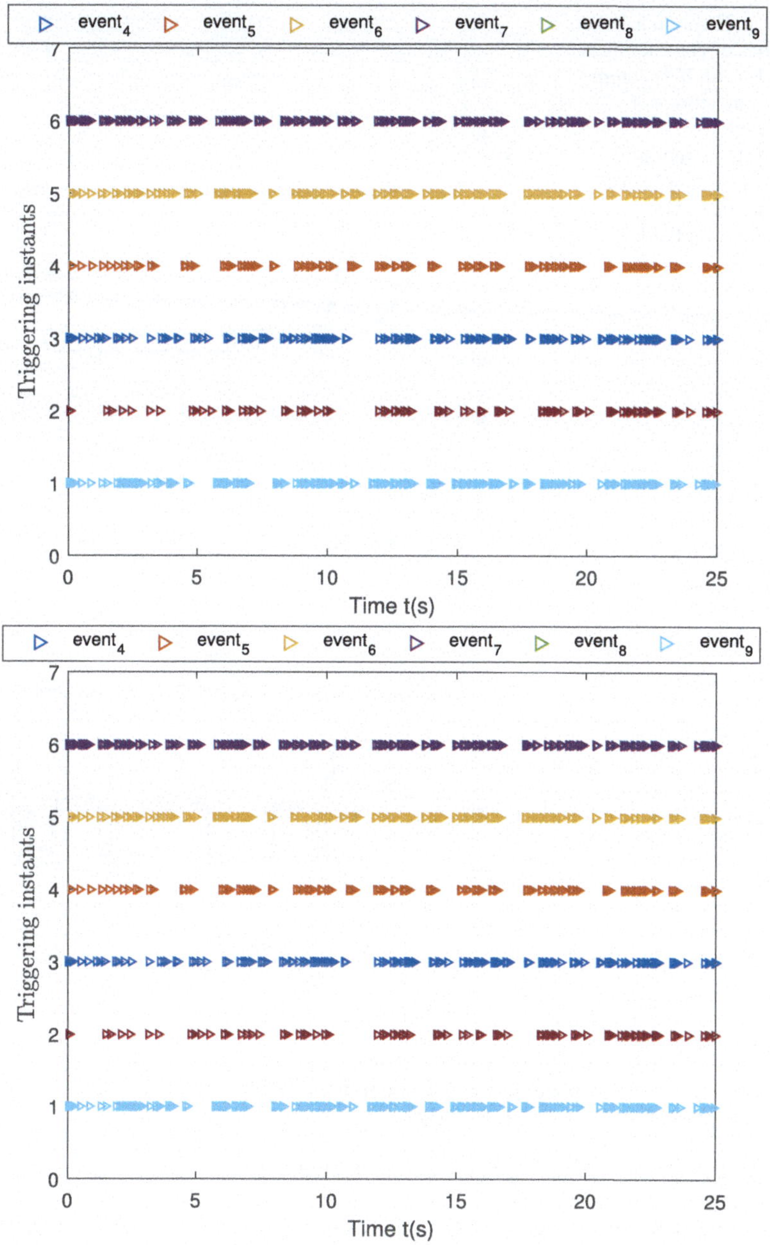

Fig. 5.5 **a** Theorem 5.1: The triggering instants t_k^i $(k \in \mathbb{N},\ i \in I[1, 6])$, **b** Theorem 5.2: The triggering instants t_k^i $(k \in \mathbb{N},\ i \in I[1, 6])$, reprinted from Zhang et al. (2023), copyright 2024, with permission from IEEE

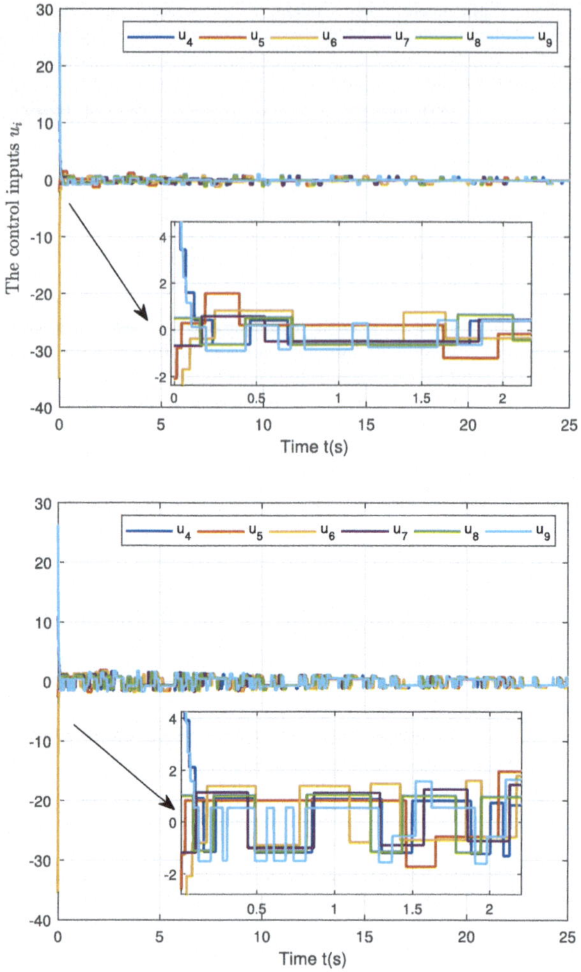

Fig. 5.6 **a** Theorem 5.1: The trajectories of control inputs $u_i,\ i \in I[1,6]$, **b** Theorem 5.2: The trajectories of control inputs $u_i,\ i \in I[1,6]$, reprinted from Zhang et al. (2023), copyright 2024, with permission from IEEE

5.1. Some relative parameters are selected as $\varsigma_1 = 0.01,\ \varsigma_2 = 1,\ \varsigma_3 = 1,\ \theta_1 = \theta_2 = \theta_3 = 1,\ \iota_1 = \iota_2 = \iota_3 = 0.05$ and $\nu = 0.11$ in Eq.(5.23), and $\vartheta(t) = \theta_1 e^{-\iota_1 t},\ \theta(t) = \theta_2 e^{-\iota_2 t},\ \chi(t) = \theta_3 e^{-\iota_3 t}$ in Li et al. (2021).

The initial states $x_i(0),\ i \in I[1,7]$ are selected as

$$x_1(0) = \mathrm{col}(0.2982, 0.0464, 0.5054),\ x_2(0) = \mathrm{col}(0.7614, 0.6311, 0.0899),$$

$$x_3(0) = \mathrm{col}(0.0809, 0.7772, 0.9051),\ x_4(0) = \mathrm{col}(0.5338, 0.1092, 0.8258),$$

$$x_5(0) = \mathrm{col}(0.3381, 0.2940, 0.7463),\ x_6(0) = \mathrm{col}(0.0103, 0.0484, 0.6679),$$

$$x_7(0) = \mathrm{col}(0.6035, 0.5261, 0.7297),$$

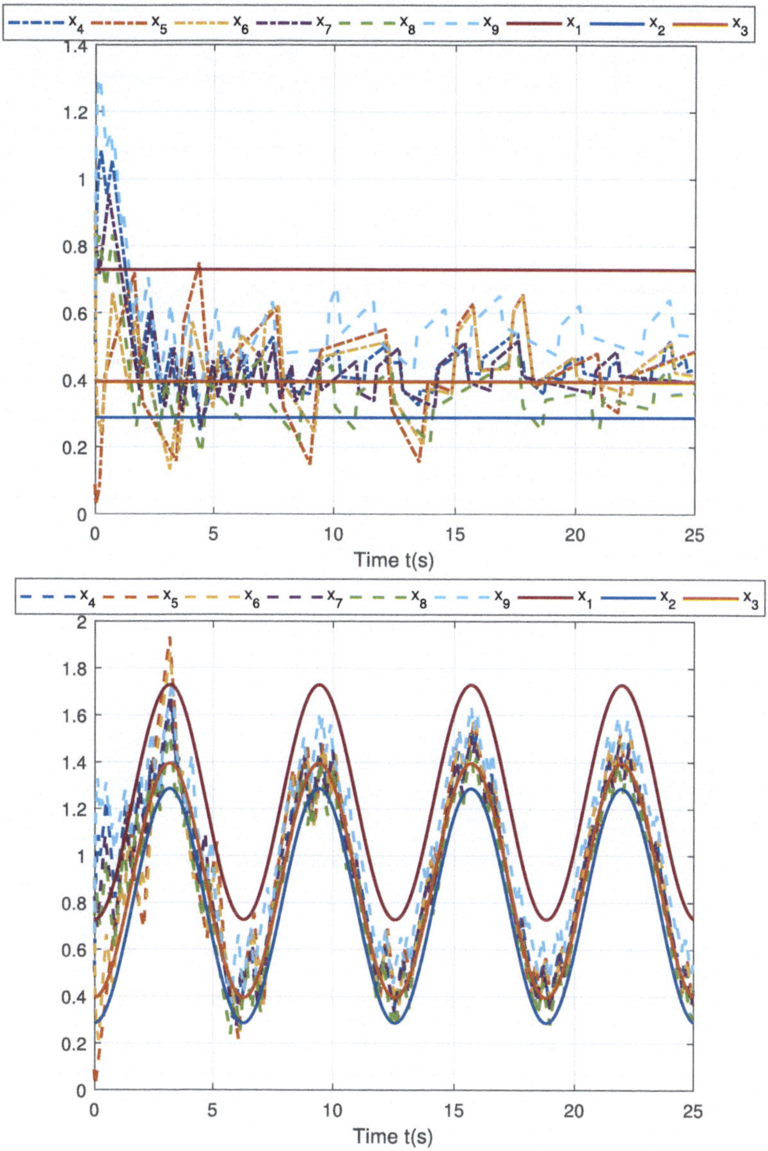

Fig. 5.7 **a** Theorem 5.1: The trajectories of states $x_{i3},\ i \in I[1, 6]$. **b** Theorem 5.2: The trajectories of states $x_{i3},\ i \in I[1, 6]$

and initial adaptive weights $c_i(0),\ i \in I[1, 7]$ are chosen as 0.001. Then, our method and the method in Li et al. (2021) were used for simulation, and the results are shown in Figs. 5.8, 5.9, and 5.10.

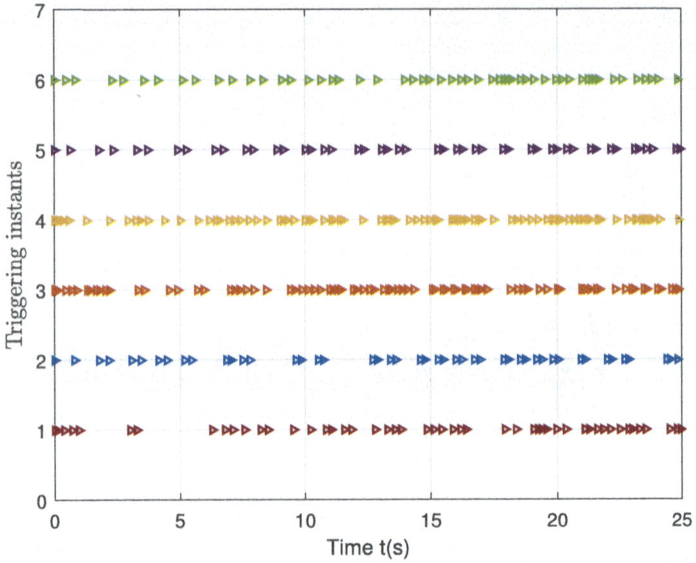

(a)

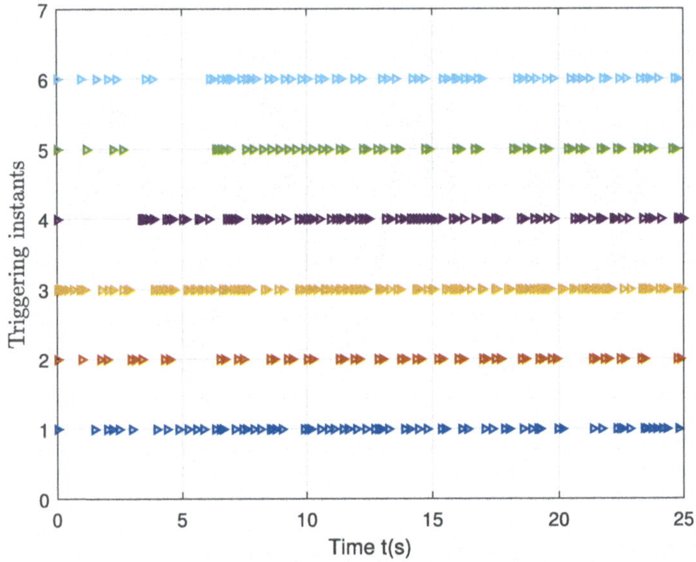

(b)

Fig. 5.8 **a** The triggering instants t_k^i $(k \in \mathbb{N},\ i \in I[1, 6])$ in this chapter, **b** The triggering instants t_k^i $(k \in \mathbb{N},\ i \in I[1, 6])$ in Li et al. (2021), reprinted from Zhang et al. (2023), copyright 2024, with permission from IEEE

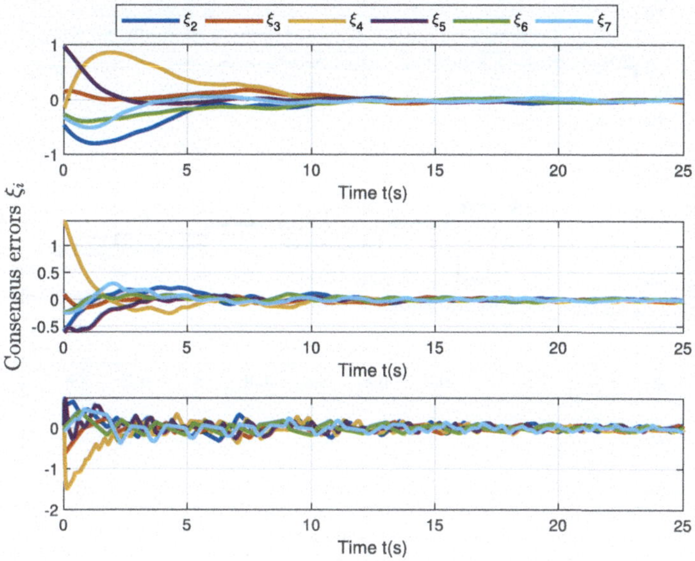

(a)

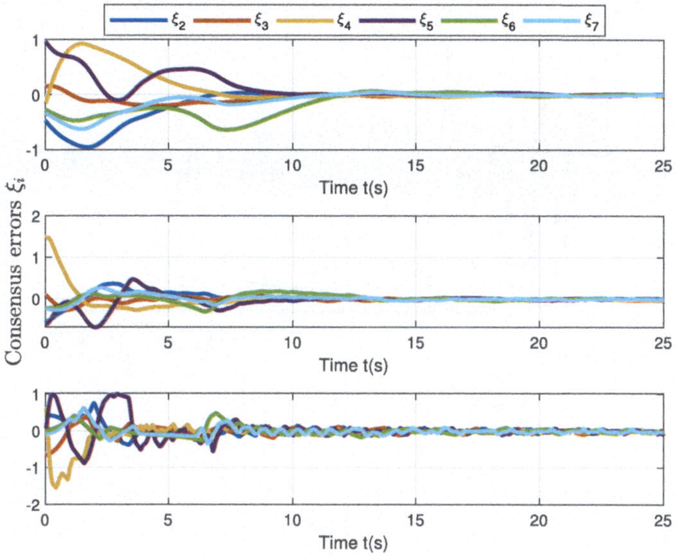

(b)

Fig. 5.9 **a** The dynamic response of error ξ_i, $i \in I[2, 7]$ in this chapter, **b** The dynamic response of error ξ_i, $i \in I[2, 7]$ in Li et al. (2021), reprinted from Zhang et al. (2023), copyright 2024, with permission from IEEE

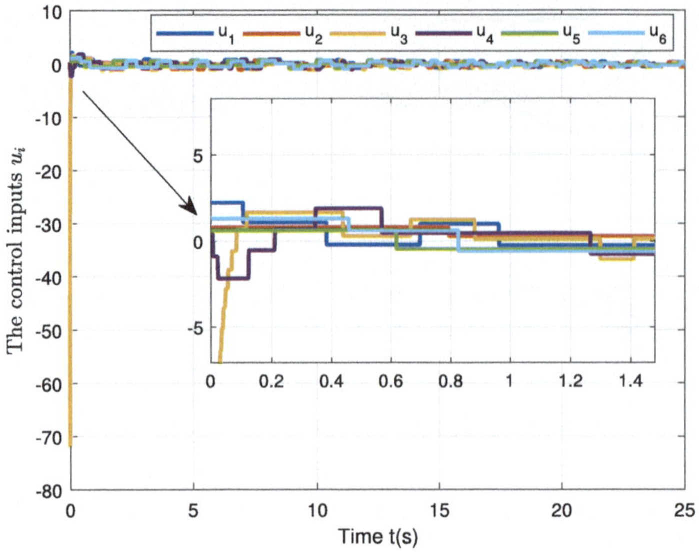

(a)

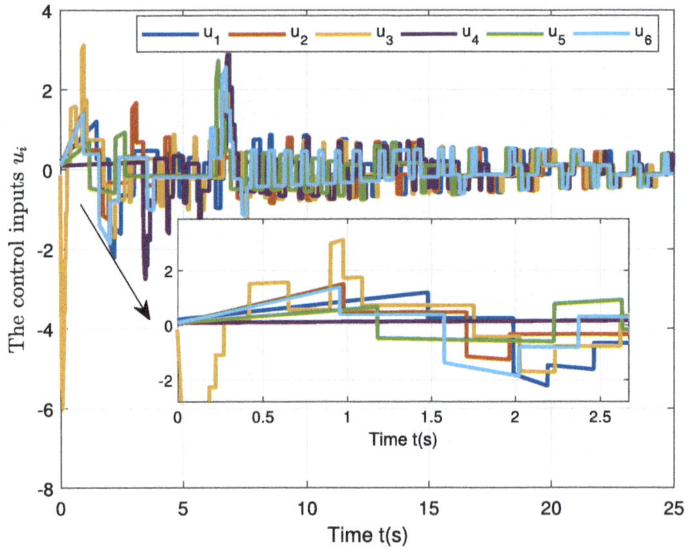

(b)

Fig. 5.10 **a** The trajectories of control inputs u_i, $i \in I[1, 6]$ in this chapter, **b** The trajectories of control inputs u_i, $i \in I[1, 6]$ in Li et al. (2021), reprinted from Zhang et al. (2023), copyright 2024, with permission from IEEE

Figures 5.8, 5.9, and 5.10 demonstrate that the method proposed in this chapter and the method in Li et al. (2021) can realize the cooperative control problem and ensure

Table 5.1 The number of triggers in Example 5.2, reprinted from Zhang et al. (2023), copyright 2024, with permission from IEEE

Triggering mechanisms	1	2	3	4	5	6	Total
Equation (5.23)	49	46	104	87	47	53	386
Equation (54) in Li et al. (2021)	68	52	156	92	53	62	483

event-triggered communication between neighbors and event-triggered updating of controller. As can be seen in Table 5.1, the method in this chapter can further reduce the communication frequency between agents compared with the method in Li et al. (2021). Thus, the methods in this chapter can further save system resources. In addition, Example 5.2 indicates that the approach in this chapter is more general and not only solves the problem considered in Li et al. (2021), but also achieves better system performance than the approach.

5.5 Summary

This chapter proposed the new distributed adaptive event-triggered control protocol that solved the containment problem of linear multiagent systems under directed graphs. For the case of zero input leaders, firstly, a distributed adaptive event-triggered control protocol and a composite event-triggered mechanism were constructed based on the relative information between neighbor agents. Secondly, to achieve both discrete measurement and discrete communication simultaneously, a new calculation method was proposed to calculate the measurement state. The Lyapunov stability theory analysis showed that the proposed method can effectively solve the containment problem. Finally, the proposed method was extended to the case of non-zero input leaders. The method proposed in this chapter did not rely on the knowledge of the Laplace matrix and ensured that intermittent communication between the neighbors and intermittent updating of the controller can be realized simultaneously when implementing containment problem of the linear multiagent systems.

References

Cai Y, Zhang H, Wang Y, Gao Z, He Q (2022) Adaptive bipartite fixed-time time-varying output formation-containment tracking of heterogeneous linear multiagent systems. IEEE Trans Neural Netw Learn Syst 33(9):4688–4698

Cao Y, Stuart D, Ren W, Meng Z (2011) Distributed containment control for multiple autonomous vehicles with double-integrator dynamics: Algorithms and experiments. IEEE Trans Control Syst Technol 19(4):929–938

Cheng B, Li Z (2019a) Fully distributed event-triggered protocols for linear multiagent networks. IEEE Trans Autom Control 64(4):1655–1662

Cheng B, Li Z (2019b) Coordinated tracking control with asynchronous edge-based event-triggered communications. IEEE Trans Autom Control 64(10):4321–4328
Cheng Y, Ugrinovskii V (2016) Event-triggered leader-following tracking control for multivariable multi-agent systems. Automatica 70:204–210
Ji M, Ferrari-Trecate G, Egerstedt M, Buffa A (2008) Containment control in mobile networks. IEEE Trans Autom Control 53(8):1972–1975
Li Z, Liu X, Ren W, Xie L (2013a) Distributed tracking control for linear multiagent systems with a leader of bounded unknown input. IEEE Trans Autom Control 58(2):518–523
Li Z, Ren W, Liu X, Fu M (2013b) Distributed containment control of multi-agent systems with general linear dynamics in the presensce of multiple leaders. Int J Robust Nonlinear Control 23:534–547
Li Z, Duan Z, Ren W, Feng G (2015) Containment control of linear multi-agent systems with multiple leaders of bounded inputs using distributed continuous controllers. Int J Robust Nonlinear Control 25:2101–2121
Li X, Sun Z, Tang Y, Karimi HR (2021) Adaptive event-triggered consensus of multiagent systems on directed graphs. IEEE Trans Autom Control 66(4):1670–1685
Liu K, Ji Z, Xu R (2016) Event-based broadcasting containment control for multi-agent systems under directed topology. Int J Control 89(11):2360–2370
Li W, Zhang H, Zhou Y, Wang Y (2022) Bipartite formation tracking for multi-agent systems using fully distributed dynamic edge-event-triggered protocol. IEEE/CAA J Autom Sinica 1–7
Wang D, Wang Z, Wang Z, Wang W (2021) Design of hybrid event-triggered containment controllers for homogeneous and heterogeneous multiagent systems. IEEE Trans Cybern 51(10):4885–4896
Xiong G, Wen G, Peng Z, Huang T (2022) Pull-based event-triggered containment control for multiagent systems with active leaders via aperiodic sampled-data transmission. IEEE Trans Syst Man Cybern Syst 52(1):214–225
Yang R, Zhang H, Feng G, Yan H (2018) Distributed event-triggered adaptive control for cooperative output regulation of heterogeneous multiagent systems under switching topology. IEEE Trans Neural Netw Learn Syst 29(9):4347–4358
Yang J, Xiao F, Ma J (2019) Model-based edge-event-triggered containment control under directed topologies. IEEE Trans Cybern 49(7):2556–2567
Zhang H, Cai Y, Wang Y, Su H (2020) Adaptive bipartite event-triggered output consensus of heterogeneous linear multiagent systems under fixed and switching topologies. IEEE Trans Neural Netw Learn Syst 31(11):4816–4830
Zhang J, Zhang H, Cai Y, Li W (2021) Containment control of general linear multi-agent systems by event-triggered control mechanisms. Neurocomputing 433(14):263–274
Zhang J, Zhang H, Li W, Wang Y (2023) Adaptive event-triggered bipartite containment control of linear multiagent systems under directed topologies. IEEE Trans Control Netw Syst 10(1):516–525
Zhang J, Zhang H, Li W, Li K (2021) Distributed edge-event triggered consensus control for multi-agent sytems by edge-based asynchronous communications. Appl Math Comput 397(15):125920

Chapter 6
Robust Formation Tracking Control of Heterogeneous Multiagent Systems Under Undirected Topology

6.1 Introduction

The previous chapters only consider the related tracking control problems, ignoring the set formation when they finish tracking the target. As UAVs are becoming more and more widely used in military and civilian applications, multi-UAV formation flight control is needed to achieve cooperative reconnaissance, combat, defense, and pesticide spraying tasks in order to enable UAVs to perform better (Liu et al. 2023).

Early proposed methods in Dong and Hu (2017), Wang et al. (2019), Hu et al. (2020) are only applicable to homogeneous multiagent systems. However, in some practical applications, such as the cooperation of unmanned ground vehicles, the agents may have heterogeneous system dynamics. Therefore, the research on formation control for heterogeneous multiagent systems (Hua et al. 2018; Zuo et al. 2019; Liu et al. 2021; Jiang et al. 2019; Hua et al. 2019) is more common than that for homogeneous multiagent systems. For example, the authors in Hua et al. (2018) investigate the time-varying output formation tracking control problem for heterogeneous multiagent systems and design distributed observers to estimate convex combinations consisting of leader states. The authors in Hua et al. (2019) study the time-varying output formation tracking problem for heterogeneous multiagent systems, noting that the formation tracking problem with only one leader in Hua et al. (2019) is a special case of the multi-leader formation tracking control problems in Liang et al. (2021), and that the latter is more robust than the former. Therefore, it is more meaningful to study the formation tracking problems for heterogeneous multiagent systems with multileaders.

Due to the limited system resources, continuous communication between neighbors will waste system resources and increase the communication burden of the system. To avoid this drawback, event-triggered control with its significant advantages is naturally applied to different systems in Xie et al. (2020), Cheng et al. (2019), Zhao et al. (2022), Peng and Sun (2020), Zhang et al. (2022), Huang et al. (2020), Yang et al. (2020) to deal with different control problems. For example, the

J. Zhang et al., *Adaptive Collaborative Control of Multiagent Systems by Event-Triggered Mechanisms*, Intelligent Control and Learning Systems 14,
https://doi.org/10.1007/978-981-97-7866-9_6

controller designed in Xie et al. (2020) employs an edge-based event-triggered mechanism to avoid continuous-time communication between neighbors. The authors in Huang et al. (2020) employ event-triggered control in the design of controllers and estimators to avoid continuous-time communication and continuous-time controller updates.

The unavoidable problem of event-triggered control is how to design the triggering mechanism so that the system does not suffer from the Zeno phenomenon and increase the adjacent triggering time interval as much as possible. Recently, the dynamic event-triggered mechanism in Girard (2015) has been applied to multiagent systems in He et al. (2020), Liu and Yang (2021), Hu et al. (2020), which can obtain the larger adjacent triggering time interval than the static event-triggered mechanism. There are almost no results in solving the time-varying output formation tracking control problem for heterogeneous multiagent systems using dynamic event-triggered mechanisms. To fill the gap, this paper solves the time-varying output formation tracking problem for heterogeneous linear multiagent systems with disturbances using dynamic event-triggered mechanisms.

6.2 Problem Formulation

Consider the heterogeneous linear multiagent system composed of M followers and N leaders. The dynamics of ith follower can be modeled by

$$\begin{aligned} \dot{x}_i(t) &= A_i x_i(t) + B_i u_i(t) + D_i \omega_i(t) \\ y_i(t) &= C_i x_i(t), \quad i \in \mathcal{F} \end{aligned} \tag{6.1}$$

and kth leader can be modeled by

$$\dot{v}_k(t) = S v_k(t), \quad y_k(t) = R v_k(t), \quad k \in \mathcal{L} \tag{6.2}$$

where $x_i \in R^{n_i}$, $u_i \in R^{m_i}$, $\omega_i \in R^{q_i}$, $y_i \in R^s$ are the state, input, disturbance, and output for the follower i, $v_k \in R^p$, $y_k \in R^s$ are the state and output the leader k, respectively, $A_i \in R^{n_i \times n_i}$, $B_i \in R^{n_i \times m_i}$, $C_i \in R^{s \times n_i}$, $D_i \in R^{n_i \times q_i}$, $S \in R^{p \times p}$, $R \in R^{s \times p}$ are given system matrices with $\text{rank}(B_i) = m_i$.

To achieve the formation control, we give the time-varying formation vector $f(t) = \text{col}\Big(f_1(t), \cdots, f_M(t)\Big) \in R^{Mp}$, where $f_i(t) \in R^p$ is the piecewise-continuously differentiable vector. The desired output formation vector is given by $f_{d_i}(t) = R f_i(t)$, $i \in \mathcal{F}$.

Formation tracking problem: For the given heterogeneous multiagent system composed of the M followers in Eq. (6.1) and N leaders in Eq. (6.2), the time-varying output formation tracking problem is solved, if for any given bounded initial states, there exist positive constants β_k $(k \in \mathcal{L})$ with $\sum_{k \in I[M+1, M+N]} \beta_k = 1$ and $\varepsilon > 0 (\in R)$ such that the output formation tracking error satisfies

$$\lim_{t\to\infty}\left\| y_i(t) - \sum_{k\in I[M+1,M+N]} \beta_k y_k(t) - f_{d_i}(t)\right\| \le \varepsilon \tag{6.3}$$

where the output formation tracking error $\epsilon_i(t) = y_i(t) - \sum_{I[M+1,M+N]} \beta_k y_k(t) - f_{d_i}(t)$ is given.

To solve the time-varying output formation tracking problem of multiagent system consisting of Eqs. (6.1) and (6.2), the following assumptions are given.

Assumption 6.1 The subgraph $\mathcal{G}$ is undirected and connected. Every follower is either well-informed or not well-informed, where a well-informed follower means that the follower has direct access to all leader's information; otherwise, the follower is not well-informed one.

Assumption 6.2 The pairs (X_i, U_i) $i \in \mathcal{F}$ are the solution of the following regulator equations:

$$\begin{aligned} X_i S &= A_i X_i + B_i U_i \\ R &= C_i X_i. \end{aligned} \tag{6.4}$$

Assumption 6.3 The pairs (A_i, B_i, C_i) are stabilizable and detectable, and the pair (R, S) is observable.

Assumption 6.4 There exists the positive constant ω such that $\|\omega_i\| \le \omega$.

Remark 6.1 It is worth emphasizing that Assumptions 6.1–6.4 are standard to deal with the formation tracking problem for heterogeneous linear multiagent systems with disturbance. Assumption 6.1 ensures that all agents can accomplish a common goal through communication. Assumptions 6.2 and 6.3 provide the necessary guarantees for the feasibility of the controller design.

6.3 Main Results

Firstly, the adaptive event-triggered distributed compensator is designed for each follower to estimate the convex combinations composed of all leaders. Secondly, based on the designed compensator, design the formation controller to realize the formation robust tracking control.

6.3.1 Adaptive Event-Triggered Distributed Compensator

Because of the presence of not well-informed followers, designing compensators based on well-informed followers and leaders designing adaptive event-triggered compensator for follower i.

$$\dot{\varsigma}_i(t) = S\varsigma_i(t) - \hat{c}_i(t)KR\Big\{ \sum_{j\in I[1,M]} a_{ij}\Big(\hat{\varsigma}_i(t) - \hat{\varsigma}_j(t)\Big) + \sum_{k\in I[M+1,M+N]} a_{ik}\Big(\hat{\varsigma}_i(t) - \upsilon_k(t)\Big)\Big\} \tag{6.5}$$

where $\varsigma_i(t) \in R^p$ denotes the compensator state, $\hat{\varsigma}_i(t)$ denotes the estimator with $\hat{\varsigma}_i(t) = \varsigma_i(t_k^i),\ t \in [t_k^i, t_{k+1}^i)$, $\hat{c}_i(t)$ denotes the adaptive weight with

$$\dot{\hat{c}}_i(t) = \mathrm{Proj}\Big(\hat{c}_i(t)\Big) = \begin{cases} 0, & \text{if } \hat{c}_i(t) \geq \bar{c}_i \\ \gamma_i q_i^T(t_k^i)\Gamma q_i(t_k^i), & \text{if } \hat{c}_i(t) < \bar{c}_i, \end{cases} \tag{6.6}$$

and $\gamma_i, \bar{c}_i$ are the given positive constants and $0 < \hat{c}_i(0) < \bar{c}_i$, $q_i(t_k^i) = \sum_{j\in I[1,M]} a_{ij}\Big(\hat{\varsigma}_i(t) - \hat{\varsigma}_j(t)\Big) + \sum_{k\in I[M+1,M+N]} a_{ik}\Big(\hat{\varsigma}_i(t) - \upsilon_k(t)\Big)$, the feedback gain matrices K, Γ are selected by $K = P^{-1}R^T$, $\Gamma = R^T R$, and $P > 0$ is the solution of the following algebraic Riccati inequality (AII):

$$S^T P + PS - R^T R + Q < 0 \tag{6.7}$$

with $Q > 0$ being the given positive definite matrix.

Remark 6.2 According to Li et al. (2011), AII (6.7) exists the solution if the following two conditions are satisfied: (1) The pair (R, S) is observable; (2) The positive definite matrix Q is set as $Q = P$.

Remark 6.3 One advantage of the compensator in Eq. (6.5) over the existing continuous-time compensator in Jiang et al. (2019) is that it reduces the demand on the network bandwidth, due to the fact that when the follower i triggers, the follower i transmits the current information to all of its neighbors. Even compared with the event-triggered compensators in Zhang et al. (2023a, b), the proposed dynamic event-triggered mechanism designed in this chapter can further reduce the triggering frequency.

Before giving the construction of the dynamic event-triggered mechanism, define the following measurement errors:

$$e_i(t) = \hat{\varsigma}_i(t) - \varsigma_i(t), \quad t \in [t_k^i, t_{k+1}^i) \tag{6.8a}$$

$$e_{q_i}(t) = q_i(t_k^i) - q_i(t),\ t \in [t_k^i, t_{k+1}^i) \tag{6.8b}$$

t_k^i ($t_0^i = 0$) is the triggering instant depending on the following dynamic event-triggered mechanism:

$$t_{k+1}^i = \inf\Big\{t > t_k^i | \theta_i\Big(e_i^T(t)\Gamma e_i(t) - \sigma_i\beta_i q_i^T(t_k^i)\Gamma q_i(t_k^i)\Big) \geq \eta_i(t)\Big\} \tag{6.9}$$

where θ_i, σ_i, β_i denote the given positive constants with $\sigma_i \in (0, 1)$, and $\eta_i(t)$ denotes the internal dynamic variable satisfying

$$\dot{\eta}_i(t) = -\rho_i \eta_i(t) + \xi_i \Big(\sigma_i \beta_i q_i^T(t_k^i) \Gamma q_i(t_k^i) - e_i^T(t) \Gamma e_i(t) \Big) \tag{6.10}$$

with $\eta_i(0) > 0$, $\rho_i > 0$, $\xi_i > 0$ being the positive constants.

Remark 6.4 It should be noted that the designed dynamic event-triggered mechanism in Eq. (6.9) avoids the need to constantly monitor the information of neighbors to determine whether the triggering condition is satisfied, although it avoids the need to continuously transmit information to neighbors. Inspired by Ding et al. (2018, 2019), this drawback can be avoided by adopting a sampled data-based event-triggered scheme. Ding *et al.* in Ding et al. (2019) applied the sampled data-based event-triggered scheme to the islanded microgrid, realizing a distributed secondary control with active power sharing and frequency regulation. However, the sampled data-based event-triggered scheme has its drawbacks, such as all agents sampling in a fixed sampling period, which is difficult to realize in large-scale networks. See Ding et al. (2018) for details.

Remark 6.5 According to Eqs. (6.9) and (6.10), we know that the design of the dynamic event-triggering mechanism contains the parameters $\theta_i, \sigma_i, \beta_i, \xi_i$, and ρ_i, $i \in$ ($\mathcal{F}$. θ_i provides degrees of freedom for the design of the trigger mechanism; the smaller θ_i, the lower the triggering frequency. $\sigma_i \beta_i$ is the threshold parameter, β_i depends on the structure of the communication topology, and the threshold can be changed by adjusting σ_i. The larger ρ_i is, the faster β_i decays.

Lemma 6.1 *For any given positive initial internal dynamic variable $\eta_i(0) > 0$, and the positive parameters $\rho_i > 0$, $\xi_i > 0$, $\theta_i > 0$, the internal dynamic variable is strictly positive and satisfies that*

$$\eta_i(t) > \eta_i(0) e^{-(\rho_i + \frac{\xi_i}{\theta_i})t} > 0, \; i \in \mathcal{F}.$$

Proof Consider the time interval $t \in \bigcup_{k=1}^{\infty} [t_k^i, t_{k+1}^i)$. According to the dynamic event-triggered mechanism in Eq. (6.9), we can know that Zeno behavior does not occur for agent i in any adjacent event interval $[t_k^i, t_{k+1}^i)$. Therefore, one obtains the following inequality:

$$\theta_i \Big(e_i^T(t) \Gamma e_i(t) - \sigma_i \beta_i q_i^T(t_k^i) \Gamma q_i(t_k^i) \Big) < \eta_i(t), \; t \in [t_k^i, t_{k+1}^i).$$

Combining the above equation and equation (6.10), we get the following inequality:

$$\dot{\eta}_i(t) \geq -\rho_i \eta_i(t) - \frac{\xi_i}{\theta_i} \eta_i(t), \; t \in [t_k^i, t_{k+1}^i).$$

Using a comparative lemma, one obtains

$$\eta_i(t) \geq \eta_i(t_k^i)e^{-(\rho_i+\frac{\xi_i}{\theta_i})(t-t_k^i)},\ t \in [t_k^i, t_{k+1}^i).$$

Based on the fact that $t_0^i = 0$, one obtains that

$$\begin{aligned}\eta_i(t) \geq &\eta_i(t_k^i)e^{-(\rho_i+\frac{\xi_i}{\theta_i})(t-t_k^i)} \geq \eta_i(t_{k-1}^i)e^{-(\rho_i+\frac{\xi_i}{\theta_i})(t-t_{k-1}^i)}\\ &\geq \cdots \geq \eta_i(0)e^{-(\rho_i+\frac{\xi_i}{\theta_i})t} > 0.\end{aligned}$$

Due to $t \in \bigcup_{k=1}^{\infty}[t_k^i, t_{k+1}^i) = [0, +\infty)$, one obtains that the internal dynamic variable satisfies $\eta_i(t) > \eta_i(0)e^{-(\rho_i+\frac{\xi_i}{\theta_i})t} > 0$ for any $t \in [0, +\infty)$. The proof is completed.

Remark 6.6 It is known from Eq. (6.10) that $\eta_i(t)$ is time-varying and its update rate depends on $q_i(t_k^i)$ and $e_i(t)$. Compared with static event-triggered mechanism, the introduction of internal variable $\eta_i(t)$ in the dynamic event-triggered mechanism is the key factor to dynamically adjust the threshold. If $\eta_i(t)$ is set to zero, the dynamic event-triggered mechanism becomes the static one in Dimarogonas et al. (2012), and the static one can be regarded as a special case of the dynamic one.

Subsequently, the time t is omitted without ambiguity. Define the following variables. $\varsigma_{\mathcal{F}} = \mathrm{col}(\varsigma_1, \cdots, \varsigma_M)$, $\upsilon_{\mathcal{L}} = \mathrm{col}(\upsilon_{M+1}, \cdots, \upsilon_{M+N})$, $\hat{c} = \mathrm{diag}\{c_1, \cdots, c_M\}$, $q(t_k) = \mathrm{col}(q_1(t_k^1), \cdots, q_M(t_k^M))$, $q = \mathrm{col}(q_1, \cdots, q_M)$ with $q_i = \sum\limits_{j \in I[1,M]} a_{ij}\Big(\varsigma_i - \varsigma_j\Big) + \sum\limits_{k \in I[M+1,M+N]} a_{ik}\Big(\varsigma_i - \upsilon_k\Big)$, $\delta = \varsigma_{\mathcal{F}} + (\mathcal{L}_1^{-1}\mathcal{L}_2 \otimes I_p)\upsilon_{\mathcal{L}}$.

According to the above definition, we can conclude that $\delta = 0$ if and only if $\varsigma_{\mathcal{F}} = -(\mathcal{L}_1^{-1}\mathcal{L}_2 \otimes I_p)\upsilon_{\mathcal{L}}$. Therefore, we call δ as the compensator error. Then, the following global form can be obtained:

$$q = (\mathcal{L}_1 \otimes I_p)\delta \tag{6.11a}$$

$$\dot{\varsigma}_{\mathcal{F}} = (I_M \otimes S)\varsigma_{\mathcal{F}} - (\hat{c} \otimes KR)q(t_k). \tag{6.11b}$$

Then, calculate the derivative of δ

$$\dot{\delta} = (I_M \otimes S)\delta - (\hat{c} \otimes KR)q(t_k). \tag{6.12}$$

A theorem is presented below to illustrate the feasibility of the proposed dynamic adaptive event-triggered compensator in Eq. (6.5).

Theorem 6.1 *Under the condition of Assumptions 6.1 and 6.3, the dynamic adaptive event-triggered compensator in Eq.* (6.5) *under the triggering mechanism in Eq.* (6.9) *is feasible, if the feedback matrices are selected as* $K = P^{-1}R^T$, $\Gamma = R^T R$, *and relative parameters are selected as*

$$\gamma_{i0} \geq \xi_i > 0,\ \rho_i > \frac{\gamma_{i0} - \xi_i}{\theta_i} \geq 0,\ \beta_i = \frac{1}{\gamma_{i0}}$$

with $\gamma_{i0} = M\Big((2c + \bar{c}_i)\Delta_{ii}^2 + \sum_{j\in I[1,M], j\neq i} a_{ji}^2(2c + \bar{c}_j)\Big)$ *and* $c > \frac{1}{\lambda_1(\mathcal{L}_1)}$, *namely, the compensator error* δ *eventually converges to zero and the multiagent system does not exhibit Zeno behavior.*

Proof First, it is shown that the compensator error eventually converges to zero, and second, it is shown that multiagent system does not exhibit Zeno behavior.

Choose the Lyapunov function as

$$V = \delta^T(\mathcal{L}_1 \otimes P)\delta + \sum_{i\in I[1,M]} \frac{1}{2\gamma_i}\tilde{c}_i^2 + \sum_{i\in I[1,M]} \eta_i$$

where P is the solution of AII (6.7), $\tilde{c}_i = \hat{c}_i - 2c - \sigma_i$ with $c > \frac{1}{\lambda_1(\mathcal{L}_1)}$, and $\bar{c}_i \geq 2c + \sigma_i$.

For the sake of writing simplicity, label V_1 and V_2 as $V_1 = \delta^T(\mathcal{L}_1 \otimes P)\delta,\ V_2 = \sum_{i\in I[1,M]} \frac{\tilde{c}_i^2}{2\gamma_i} + \sum_{i\in I[1,M]} \eta_i$. Based on Eq. (6.11a), we can calculate the derivative of V_1 along (6.12) as

$$\begin{aligned}
\dot{V}_1 &= \delta^T\Big(\mathcal{L}_1 \otimes (PS + S^T P)\Big)\delta - 2\delta^T(\mathcal{L}_1\hat{c} \otimes \Gamma)q(t_k) \\
&= \delta^T\Big(\mathcal{L}_1 \otimes (PS + S^T P)\Big)\delta - 2q^T(\hat{c} \otimes \Gamma)q(t_k) \\
&= \delta^T\Big(\mathcal{L}_1 \otimes (PS + S^T P)\Big)\delta - cq^T(I_M \otimes \Gamma)q \\
&+ cq^T(I_M \otimes \Gamma)q - 2q^T(\hat{c} \otimes \Gamma)q(t_k) \\
&= \delta^T\Big(\mathcal{L}_1 \otimes (PS + S^T P) - c\mathcal{L}_1\mathcal{L}_1 \otimes \Gamma\Big)\delta \\
&+ cq^T(I_M \otimes \Gamma)q - 2q^T(\hat{c} \otimes \Gamma)q(t_k).
\end{aligned} \tag{6.13}$$

In light of Eq. (6.8b) and Lemma 1.6, we can obtain that

$$\begin{aligned}
& cq^T(I_M \otimes \Gamma)q - 2q^T(\hat{c} \otimes \Gamma)q(t_k) \\
&= c\sum_{i\in I[1,M]} q_i^T\Gamma q_i - 2\sum_{i\in I[1,M]} \hat{c}_i q_i^T\Gamma q_i(t_k^i) \\
&= c\sum_{i\in I[1,M]} q_i^T\Gamma q_i - 2i \in I[1,M]\hat{c}_i q_i^T(t_k^i)\Gamma q_i(t_k^i) \\
&+ 2\sum_{i\in I[1,M]} \hat{c}_i e_{q_i}^T\Gamma q_i(t_k^i) \\
&\leq c\sum_{i\in I[1,M]} q_i^T\Gamma q_i - \sum_{i\in I[1,M]} \hat{c}_i q_i^T(t_k^i)\Gamma q_i(t_k^i)
\end{aligned}$$

$$
\begin{aligned}
&+ \sum_{i\in I[1,M]} \hat{c}_i e_{q_i}^T \Gamma e_{q_i} \\
&= \sum_{i\in I[1,M]} (c+\hat{c}_i) e_{q_i}^T \Gamma e_{q_i} - \sum_{i\in I[1,M]} (\hat{c}_i - c) q_i^T(t_k^i)\Gamma q_i(t_k^i) \\
&- 2c \sum_{i\in I[1,M]} q_i^T(t_k^i)\Gamma e_{q_i} \\
&\le \sum_{i\in I[1,M]} (2c+\hat{c}_i) e_{q_i}^T \Gamma e_{q_i} - \sum_{i\in I[1,M]} (\hat{c}_i - 2c) q_i^T(t_k^i)\Gamma q_i(t_k^i).
\end{aligned} \tag{6.14}
$$

Based on Eqs. (6.8a) and (6.8b), one obtains

$$
e_{q_i} = \sum_{i\in I[1,M]} a_{ij}\big(e_i - e_j\big) + \sum_{k\in I[M+1,N]} a_{ik} e_i .
$$

Therefore, one obtains the following inequality:

$$
\begin{aligned}
\|Re_{q_i}\| &= \|R\Big\{ \sum_{j\in I[1,M]} a_{ij}\big(e_i - e_j\big) + \sum_{k\in I[M+1,M+N]} a_{ik} e_i \Big\}\| \\
&\le \Delta_{ii}\|Re_i\| + \sum_{j\in I[1,M], j\neq i} a_{ij}\|Re_j\|,
\end{aligned} \tag{6.15}
$$

where $\Delta_{ii} = \sum_{i\in I[1,N]} a_{ij}$ is given. In addition, using $\big(\sum_{i\in I[1,M]} y_i\big)^2 \le M\sum_{i\in I[1,M]} y_i^2$, one obtains

$$
\|Re_{q_i}\|^2 \le M\Big(\Delta_{ii}^2\|Re_i\|^2 + \sum_{j\in I[1,M], j\neq i} a_{ij}^2\|Re_j\|^2\Big). \tag{6.16}
$$

Therefore, one obtains

$$
\begin{aligned}
&\sum_{i\in I[1,M]} (2c+\hat{c}_i) e_{q_i}^T \Gamma e_{q_i} = \sum_{i\in I[1,M]} (2c+\hat{c}_i)\|Re_{q_i}\|^2 \\
&\le M \sum_{i\in I[1,M]} (2c+\bar{c}_i)\Big(\Delta_{ii}^2\|Re_i\|^2 + \sum_{j=1, j\neq i}^{N} a_{ij}^2\|Re_j\|^2\Big) \\
&= M \sum_{i\in I[1,M]} \Big((2c+\bar{c}_i)\Delta_{ii}^2 + \sum_{j\in I[1,M], j\neq i} a_{ji}^2(2c+\bar{c}_j)\Big)\|Re_i\|^2 \\
&= \sum_{i\in I[1,M]} \gamma_{i0}\|Re_i\|^2,
\end{aligned} \tag{6.17}
$$

where $\gamma_{i0} = M\Big((2c+\bar{c}_i)\Delta_{ii}^2 + \sum_{j\in I[1,M], j\neq i} a_{ji}^2(2c+\bar{c}_j)\Big)$. Substituting Eqs. (6.14) and (6.17) into (6.13) yields

$$
\begin{aligned}
\dot{V}_1 \leq & \delta^T\Big(L_1 \otimes (PS + S^T P) - c\mathcal{L}_1\mathcal{L}_1 \otimes \Gamma\Big)\delta + \sum_{i\in I[1,M]} \gamma_{i0} e_i^T \Gamma e_i \\
& - \sum_{i\in I[1,M]} (\hat{c}_i - 2c) q_i^T(t_k^i)\Gamma q_i(t_k^i).
\end{aligned} \tag{6.18}
$$

Calculating the derivate of V_2 along Eq. (6.10), and combining the derivative of V_1, one obtains

$$
\begin{aligned}
\dot{V} \leq & \delta^T\Big(L_1 \otimes (PS + S^T P) - c\mathcal{L}_1\mathcal{L}_1 \otimes \Gamma\Big)\delta + \sum_{i\in I[1,M]} \gamma_{i0} e_i^T \Gamma e_i \\
& - \sum_{i\in I[1,M]} (\hat{c}_i - 2c) q_i^T(t_k^i)\Gamma q_i(t_k^i) \\
& + \sum_{i\in I[1,M]} \Big(- \rho_i\eta_i + \xi_i\big(\sigma_i\beta_i q_i^T(t_k^i)\Gamma q_i(t_k^i) - e_i^T\Gamma e_i\big)\Big) \\
& + \sum_{i\in I[1,M]} \frac{1}{\gamma_i}\tilde{c}_i\dot{\tilde{c}}_i.
\end{aligned} \tag{6.19}
$$

Taking advantage of $\beta_i = \frac{1}{\gamma_{i0}}$ and $\tilde{c}_i = \hat{c}_i - 2c - \sigma_i$, one can obtain

$$
\begin{aligned}
\dot{V} \leq & \delta^T\Big(L_1 \otimes (PS + S^T P) - c\mathcal{L}_1\mathcal{L}_1 \otimes \Gamma\Big)\delta \\
& + \sum_{i\in I[1,M]} (\gamma_{i0} - \xi_i) e_i^T \Gamma e_i - \sum_{i\in I[1,M]} \tilde{c}_i q_i^T(t_k^i)\Gamma q_i(t_k^i) \\
& - \sum_{i\in I[1,M]} (\gamma_{i0} - \xi_i)\sigma_i\beta_i q_i^T(t_k^i)\Gamma q_i(t_k^i) \\
& - \sum_{i\in I[1,M]} \rho_i\eta_i + \sum_{i\in I[1,M]} \frac{1}{\gamma_i}\tilde{c}_i\dot{\tilde{c}}_i.
\end{aligned} \tag{6.20}
$$

Using the dynamic event-triggered mechanism in Eq. (6.9), one can obtain Eq. (6.20) satisfies

$$
\begin{aligned}
\dot{V} \leq & \delta^T\Big(L_1 \otimes (PS + S^T P) - c\mathcal{L}_1\mathcal{L}_1 \otimes \Gamma\Big)\delta \\
& + \sum_{i\in I[1,M]} \Big(\frac{\gamma_{i0} - \xi_i}{\theta_i} - \rho_i\Big)\eta_i - \sum_{i\in I[1,M]} \tilde{c}_i q_i^T(t_k^i)\Gamma q_i(t_k^i) \\
& + \sum_{i\in I[1,M]} \frac{1}{\gamma_i}\tilde{c}_i\dot{\tilde{c}}_i.
\end{aligned} \tag{6.21}
$$

Discuss in two cases for the formula $-\sum_{i\in I[1,M]} \tilde{c}_i q_i^T(t_k^i)\Gamma q_i(t_k^i) + \sum_{i\in I[1,M]} \frac{1}{\gamma_i}\tilde{c}_i\dot{\tilde{c}}_i$.

Case 1: If $\hat{c}_i \geq \bar{c}_i$, it is easy see that $\dot{\tilde{c}}_i = 0$. Therefore,

$$- \sum_{i\in I[1,M]} \tilde{c}_i q_i^T(t_k^i)\Gamma q_i(t_k^i) + \sum_{i\in I[1,M]} \frac{1}{\gamma_i}\tilde{c}_i\dot{\tilde{c}}_i \le 0.$$

Case 2: If $\hat{c}_i < \bar{c}_i$, it is easy see that $\dot{\tilde{c}}_i = \gamma_i q_i^T(t_k^i)\Gamma q_i(t_k^i)$. Therefore,

$$- \sum_{i\in I[1,M]} \tilde{c}_i q_i^T(t_k^i)\Gamma q_i(t_k^i) + \sum_{i\in I[1,M]} \frac{1}{\gamma_i}\tilde{c}_i\dot{\tilde{c}}_i = 0.$$

Combining Case 1 and Case 2, we can derive the following formula:

$$- \sum_{i\in I[1,M]} \tilde{c}_i q_i^T(t_k^i)\Gamma q_i(t_k^i) + \sum_{i\in I[1,M]} \frac{1}{\gamma_i}\tilde{c}_i\dot{\tilde{c}}_i \le 0. \tag{6.22}$$

Substituting Eqs. (6.22) into (6.21) yields

$$\begin{aligned}\dot{V} \le & \delta^T\Big(L_1 \otimes (PS + S^T P) - c\mathcal{L}_1\mathcal{L}_1 \otimes \Gamma\Big)\delta \\ & + \sum_{i\in I[1,M]} (\frac{\gamma_{i0} - \xi_i}{\theta_i} - \rho_i)\eta_i.\end{aligned} \tag{6.23}$$

Since $\mathcal{L}_1$ is the positive definite matrix, there exists the orthogonal matrix Λ satisfying $\Lambda^T\mathcal{L}_1\Lambda = \mathrm{diag}\{\lambda_1(\mathcal{L}_1), \cdots, \lambda_M(\mathcal{L}_1)\}$. Let $\hat{\delta} = (\Lambda^T \otimes I_p)\delta$, submit it into Eq. (6.23) and use $c > \frac{1}{\lambda_1(\mathcal{L}_1)}$ to simplify

$$\begin{aligned}\dot{V} \le & \sum_{i\in I[1,M]} \hat{\delta}_i^T\Big(\lambda_i(\mathcal{L}_1)(PS + S^T P) - c\lambda_i^2(\mathcal{L}_1)\Gamma\Big)\hat{\delta}_i \\ & + \sum_{i\in I[1,M]} (\frac{\gamma_{i0} - \xi_i}{\theta_i} - \rho_i)\eta_i \\ \le & - \sum_{i\in I[1,M]} \lambda_i(\mathcal{L}_1)\hat{\delta}_i^T Q\hat{\delta}_i + \sum_{i\in I[1,M]} (\frac{\gamma_{i0} - \xi_i}{\theta_i} - \rho_i)\eta_i.\end{aligned} \tag{6.24}$$

According the fact that $\rho_i - \frac{\gamma_{i0}-\xi_i}{\theta_i} > 0$, we obtain $\dot{V} \le 0$.

Let $\eta = \mathrm{col}(\eta_1, \cdots, \eta_M)$, $\tilde{c} = \mathrm{col}(\tilde{c}_1, \cdots, \tilde{c}_M)$. It is not difficult to obtain that $\dot{V} = 0$ if and only if $\delta = 0$ and $\eta = 0$, $i \in \mathcal{F}$. Thus, the set $\mathbb{E} = \{(\delta, \eta, \tilde{c})|\dot{V} = 0\}$ is the largest invariant set with $\{\dot{V} = 0\}$. According to LaSalle's invariant principle, one obtains that $\lim_{t\to\infty}\delta = 0$, $\lim_{t\to\infty}\eta = 0$, and $\tilde{c}$ eventually converges to the finite value. Summarizing the above analysis, we can conclude that the proposed compensator in Eq. (6.5) is feasible.

Finally, we analyze that the proposed dynamic event-triggered mechanism in Eq. (6.9) is feasible, that is, each subsystem will not exhibit Zeno behavior. Based on Eqs. (6.5) and (6.8a), calculate the right-hand derivative of e_i as $D^+e_i = -S\varsigma_i + \hat{c}_i K R q_i(t_k^i)$, $t \in [t_k^i, t_{k+1}^i)$. Then, one can obtain the following inequality:

$$
\begin{aligned}
D^{+}\|e_i\| \leq & \| - S\varsigma_i + \hat{c}_i K R q_i(t_k^i)\| \\
\leq & \|S\|\|e_i\| + \|S\|\|\varsigma\| + \hat{c}_i\|K R q_i(t_k^i)\| \\
= & \|S\|\|e_i\| + \Psi_k^{ij},
\end{aligned} \tag{6.25}
$$

where Ψ_k^{ij} denotes the upper bound of $\|S\|\|\hat{\varsigma}\| + \hat{c}_i\|K R q_i(t_k^i)\|$.

Subsequently, the following two scenarios are used to exclude Zeno behavior.

Case 1: If $\|S\| \neq 0$. According to Eq. (6.25) and comparison lemma, e_i satisfies

$$
\|e_i\| \leq \frac{\Psi_k^{ij}}{\|S\|}\left(e^{\|S\|(t-t_k^i)} - 1\right).
$$

Based on the above inequality, we can obtain

$$
\|Re_i\| \leq \frac{\Psi_k^{ij}\|R\|}{\|S\|}\left(e^{\|S\|(t-t_k^i)} - 1\right). \tag{6.26}
$$

According to the dynamic event-triggered mechanism in Eq. (6.9) and Lemma 6.1, one obtains

$$
\begin{aligned}
\|Re_i(t_{k+1}^i)\|^2 \geq & \sigma_i\beta_i\|Rq_i(t_k^i)\|^2 + \frac{\eta_i(t_{k+1}^i)}{\theta_i} \\
\geq & \sigma_i\beta_i\|Rq_i(t_k^i)\|^2 + \frac{\eta_i(0)}{\theta_i}e^{-(\rho_i+\frac{\xi_i}{\theta_i})t_{k+1}^i}.
\end{aligned} \tag{6.27}
$$

Combining Eqs. (6.26) and (6.27) yields

$$
\begin{aligned}
0 \leq & \frac{\eta_i(0)}{\theta_i}e^{-(\rho_i+\frac{\xi_i}{\theta_i})t_{k+1}^i} \leq \sigma_i\beta_i\|Rq_i(t_k^i)\|^2 + \frac{\eta_i(0)}{\theta_i}e^{-(\rho_i+\frac{\xi_i}{\theta_i})t_{k+1}^i} \\
\leq & \left\{\frac{\Psi_{k+1}^{ij}\|R\|}{\|S\|}\left(e^{\|S\|(t_{k+1}^i-t_k^i)} - 1\right)\right\}^2.
\end{aligned} \tag{6.28}
$$

It follows from Eq. (6.28) that

$$
t_{k+1}^i - t_k^i \geq \frac{1}{\|S\|}\ln\left\{1 + \frac{\|S\|}{\Psi_{k+1}^{ij}\|R\|}\sqrt{\sigma_i\beta_i\|Rq_i(t_k^i)\|^2 + \frac{\eta_i(0)}{\theta_i}e^{-(\beta+\frac{\xi_i}{\theta_i})t_{k+1}^i}}\right\}. \tag{6.29}
$$

Therefore, by the definition of excluding Zeno behavior we know that it is only necessary to analyze the above inequality as strictly positive. Here we use the contradiction argument approach, assuming that Zeno behavior occurs in the agent i, $\lim\limits_{k\to\infty} t_{k+1}^i - t_k^i = 0$. According to Eq. (6.28), we can obtain

$$
\lim_{\kappa\to\infty}\frac{\eta_i(0)}{\theta_i}e^{-(\rho_i+\frac{\xi_i}{\theta_i})t_{k+1}^i} \leq \lim_{\kappa\to\infty}\left\{\frac{\Psi_{k+1}^{ij}\|R\|}{\|S\|}\left(e^{\|S\|(t_{k+1}^i-t_k^i)} - 1\right)\right\}^2.
$$

Therefore, one has

$$
\begin{aligned}
&\lim_{\kappa\to\infty} \frac{\eta_i(0)}{\theta_i} e^{-(\rho_i+\frac{\xi_i}{\theta_i})(t_{k+1}^i-t_k^i)} \\
&\le \lim_{\kappa\to\infty} e^{(\rho_i+\frac{\xi_i}{\theta_i})t_k^i}\left\{\frac{\Psi_{k+1}^{ij}\|R\|}{\|S\|}\left(e^{\|S\|(t_{k+1}^i-t_k^i)}-1\right)\right\}^2.
\end{aligned}
\tag{6.30}
$$

Combining $\lim_{k\to\infty} t_{k+1}^i - t_k^i = 0$ and Eq. (6.30), we can conclude that $\frac{\eta_i(0)}{\theta_i} \le 0$, which is contradictory to the fact $\frac{\eta_i(0)}{\theta_i} > 0$. Therefore, the assumption is wrong and Zeno behavior does not occur in this case.

Case 2: If $\|S\| = 0$. According to Eq. (6.25), one can obtain that

$$
\|Re_i\| \le \Psi_k^{ij}\|R\|(t-t_k^i),\ t\in[t_k^i, t_{k+1}^i). \tag{6.31}
$$

According to Eqs. (6.27) and (6.31), one obtains that

$$
\begin{aligned}
0 \le &\frac{\eta_i(0)}{\theta_i} e^{-(\rho_i+\frac{\xi_i}{\theta_i})t_{k+1}^i} \le \sigma_i\beta_i\|Rq_i(t_k^i)\|^2 + \frac{\eta_i(0)}{\theta_i} e^{-(\rho_i+\frac{\xi_i}{\theta_i})t_{k+1}^i} \\
\le &\Psi_{k+1}^{ij}\|R\|(t_{k+1}^i - t_k^i).
\end{aligned}
\tag{6.32}
$$

The rest of the proof is similar to the situation in Case 1 and is omitted here for reasons of space. The proof is completed.

6.3.2 Formation Controller

In order for the multiagent system to form a predetermined formation and follow the virtual leader, the formation tracking controller for agent i is designed as

$$
\begin{aligned}
u_i &= K_{x_i}\chi_i + K_{\varsigma_i}(\varsigma_i + f_i) + r_i \\
\dot{\chi}_i &= A_i\chi_i + B_i u_i + G_i(C_i\chi_i - y_i)
\end{aligned}
\tag{6.33}
$$

where $\chi_i \in R^{n_i}$ denotes the observer state, $r_i \in R^{m_i}$ denotes the time-varying compensation input, f_i denotes the time-varying vector, and $K_{x_i} \in R^{m_i\times n_i}$, $K_{\varsigma_i} \in R^{m_i\times p}$, and $G_i \in R^{n_i\times s}$ denote the feedback gain matrices given in Algorithm 1 to be designed.

Algorithm 6.1 Design of relevant parameters of the formation controller in Eq. (6.33).

Step 1: Select the matrices $\hat{B}_i \in R^{m_i\times n_i}$ and $\tilde{B}_i \in R^{(n_i-m_i)\times n_i}$ such that $\hat{B}_i B_i = I_{m_i}$ and $\tilde{B}_i B_i = 0$. Moreover, the matrix $\Xi_i = \mathrm{col}(\hat{B}_i, \tilde{B}_i)$ is invertible.

Step 2: Select the time-varying formation vector $f_{d_i} = Rf_i$ with

$$\lim_{t\to\infty}\left(\tilde{B}_i X_i (Sf_i - \dot{f}_i)\right) = 0. \tag{6.34}$$

Step 3: Select the formation compensator input r_i $(i \in \mathcal{F})$ with

$$r_i = -\hat{B}_i X_i (Sf_i - \dot{f}_i). \tag{6.35}$$

Step 4: Select K_{x_i} and G_i with $A_i + B_i K_{x_i}$ and $A_i + G_i C_i$ being Hurwitz, and select the matrix $K_{\varsigma i}$ with $K_{\varsigma i} = U_i - K_{x_i} X_i$.

Remark 6.7 From Eq. (6.34) it is known that the formation vector has the constraint, which is common, where compensation input in (6.35) is used to extend the set of feasible formation. Similar assumptions can be found in Zuo et al. (2019), Jiang et al. (2019), Zhang et al. (2023a), Ge and Han (2017). It is worth emphasizing that the formation vector f_{di} is designed based on the actual task requirements, f_i can be further obtained by $f_{d_i} = Rf_i$, and must satisfy the formation feasibility condition in Eq. (6.34).

Theorem 6.2 *Under the conditions of Assumption 4.1–4.4, the time-varying formation vector f_{d_i} and the compensation input r_i are designed in Algorithm 6.1, and the compensator designed in Eq. (6.5), the time-varying output formation tracking problem of multiagent system composed of Eqs. (6.1) and (6.2) can be solved under the formation tracking controller in Eq. (6.33) with relative parameters given in Algorithm 6.1.*

Proof Set

$$e_{x_i} = \chi_i - x_i$$

$$\varpi = \frac{1}{\sum_{j\in I[M+1,M+N]} a_j} \sum_{k\in I[M+1,M+N]} a_k v_k$$

$$\delta_i = \varsigma_i - \varpi$$

$$\zeta_i = x_i - X_i(\varpi + f_i)$$

where a_j satisfies $-L_1^{-1}L_2 = 1_M \otimes \frac{[a_{M+1},\cdots,a_{M+N}]}{\sum_{j\in I[M+1,M+N]} a_j}$.

Calculate the derivative of e_{x_i} as

$$\dot{e}_{x_i} = (A_i + G_i C_i)e_{x_i} - D_i \omega_i. \tag{6.36}$$

According to Eq. (6.36), one can obtain that

$$e_{x_i} = e^{(A_i+G_iC_i)(t-t_0)}e_{x_i}(t_0) + \int_{t_0}^{t} e^{(A_i+G_iC_i)(t-\tau)}D_i\omega_i(\tau)d\tau.$$

Based on the fact that $A + G_iC_i$ is the Hurwitz matrix, one can obtain $\|e^{(A_i+G_iC_i)(t-t_0)}\| \leq k_1 e^{-\lambda_1(t-t_0)}$ with $k_1 > 0$ and $\lambda_1 > 0$. Further, we have

$$\|e_{x_i}\| \leq k_1 e^{-\lambda_1(t-t_0)}\|e_{x_i}(t_0)\| + \frac{k_1\|D_i\|}{\lambda_1}\sup_{t_0\leq\tau\leq t}\|\omega_i\|. \tag{6.37}$$

Therefore, we obtain $\lim_{t\to\infty}\|e_{x_i}\| \leq \frac{k_1\|D_i\|}{\lambda_1}\omega$.

Based on the formation tracking controller in (6.33), we can obtain the dynamics of the closed-loop system satisfies that

$$\begin{aligned}\dot{x}_i &= A_i x_i + B_i\Big(K_{x_i}\chi_i + K_{\varsigma_i}(\varsigma_i + f_i) + r_i\Big)\\ &= (A_i + B_iK_{x_i})x_i + B_iK_{x_i}e_{x_i} + D_i\omega_i + B_i(U_i - K_{x_i}X_i)(\varsigma_i + f_i) + B_ir_i.\end{aligned} \tag{6.38}$$

Further, calculate the derivative of ζ_i,

$$\begin{aligned}\dot{\zeta}_i &= (A_i + B_iK_{x_i})\Big(\zeta_i + X_i(s_i\varpi + f_i)\Big) + B_iK_{x_i}e_{x_i}\\ &\quad + B_i(U_i - K_{x_i}X_i)(\delta_i + s_i\varpi + f_i)\\ &\quad + B_ir_i - X_is_i\dot{\varpi} - X_i\dot{f}_i + D_i\omega_i\\ &= (A_i + B_iK_{x_i})\zeta_i + (A_iX_i + B_iU_i)s_i\varpi\\ &\quad + B_iK_{x_i}e_{x_i} + B_iK_{\varsigma_i}\delta_i + B_ir_i\\ &\quad - X_is_i\dot{\varpi} - X_i\dot{f}_i + D_i\omega_i + X_iSf_i.\end{aligned} \tag{6.39}$$

Based on Eq. (6.2), we obtain that $\dot{\varpi} = S\varpi$. Then, substitute $\dot{\varpi} = S\varpi$ into Eq. (6.39), and take advantage of Eq. (6.4) yields

$$\dot{\zeta}_i = (A_i + B_iK_{x_i})\zeta_i + B_iK_{x_i}e_{xi} + B_iK_{\varsigma_i}\delta_i + X_i(Sf_i - \dot{f}_i) + B_ir_i + D_i\omega_i. \tag{6.40}$$

According to the fact that $\hat{B}_iB_i = I_{m_i}$ and Eq. (6.35), one obtains

$$\lim_{t\to\infty}\Big(\tilde{B}_iX_i(Sf_i - \dot{f}_i) + \tilde{B}_iB_ir_i\Big) = 0. \tag{6.41}$$

Further, based on the compensation input r_i, we have

$$\hat{B}_iX_i(Sf_i - \dot{f}_i) + \hat{B}_iB_ir_i = 0. \tag{6.42}$$

Combining the two equations above, we obtain

$$\lim_{t\to\infty} \left(X_i(Sf_i - \dot{f}_i) + B_i r_i\right) = 0. \tag{6.43}$$

It follows from Theorem 6.1 that $\lim_{t\to\infty} \delta = 0$. In addition, we obtain that $A_i + B_i K_{x_i}$ is Hurwitz and $\lim_{t\to\infty} \|e_{x_i}\| \leq \frac{k_1\|D_i\|}{\lambda_1}\omega$. Similar to Eq. (6.37), it is easy to get $\lim_{t\to\infty} \|\zeta_i\| \leq \frac{k_2}{\lambda_2}(\frac{\|B_1 K_{x_i}\|k_1}{\lambda_1} + 1)\|D_i\|\omega$ with $k_2, \lambda_2 > 0$, namely,

$$\lim_{t\to\infty} \|x_i - X_i(s_i\varpi + f_i)\| \leq \frac{k_2}{\lambda_2}(\frac{\|B_1 K_{x_i}\|k_1}{\lambda_1} + 1)\|D_i\|\omega.$$

Set $\varepsilon = \max_{i\in\mathcal{F}} \left\{\|C_i\|\frac{k_2}{\lambda_2}(\frac{\|B_1 K_{x_i}\|k_1}{\lambda_1} + 1)\|D_i\|\omega\right\}$. Then, the above formula can be written as

$$\lim_{t\to\infty} \|y_i - s_i R\varpi - f_{d_i}\| \leq \varepsilon \tag{6.44}$$

where $\varepsilon = \max_{i\in\mathcal{F}} \left\{\|C_i\|\frac{k_2}{\lambda_2}(\frac{\|B_1 K_{x_i}\|k_1}{\lambda_1} + 1)\|D_i\|\omega\right\}$. Further, we can obtain that

$$\lim_{t\to\infty} \left\|y_i - \sum_{k\in I[M+1,M+N]} \beta_k y_k - f_{d_i}\right\| \leq \varepsilon \tag{6.45}$$

where $\beta_k = \frac{a_k}{\sum_{j\in I[M+1,N]} a_j}$ is given and satisfies that $\sum_{k\in I[M+1,N]} \beta_k = 1$. Obviously, based on the definition of the formation tracking problem, we can know that the multiagent system solves the formation tracking problem under the action of the proposed dynamic event-triggered compensator in Eq. (6.5) and formation tracking controller in Eq. (6.33). The proof is completed.

Remark 6.8 From Theorem 6.2, it is known that the formation tracking error eventually converges to an adjustable bound ε, ε depending on the size of the disturbance. The smaller the disturbance, the smaller ε is. When the disturbance is absent, the formation tracking error ϵ_i converges to 0, i.e., asymptotic convergence is obtained in this case.

Remark 6.9 Compared with the existing related work, the advantages of the approach in this chapter are: 1) Comparing with the work in Dong et al. (2017), Liang et al. (2023), Yan et al. (2022), even Li et al. (2022), this chapter introduces the dynamic event-triggered mechanism, which significantly reduces the communication frequency between neighbors and saves system communication resources. 2) The multiagent system considered in this chapter is heterogeneous and consists of multiple leaders, and the scenarios considered in Dong et al. (2017), Yan et al. (2022), Li et al. (2022) can be viewed as special cases of this chapter.

6.4 Simulation Examples

In order to verify the feasibility of the methods in this chapter, two examples are given in this section to aid in verification.

Example 6.1 Consider the heterogeneous linear multiagent system consisting of Eqs. (6.1) and (6.2), where the system matrices are given by

$$A_1 = A_2 = I_2 \otimes \begin{bmatrix} 0\ 1; 0\ 0 \end{bmatrix},\ A_3 = 0_{2\times 2},\ A_4 = A_5 = I_2 \otimes \begin{bmatrix} 0 & 1 & 0 \\ 0 & 0 & 1 \\ -1 & -2 & -1 \end{bmatrix},$$

$$A_6 = A_7 = I_2 \otimes \begin{bmatrix} 0 & 1 & 0 \\ 0 & 0 & 1 \\ 2 & -2 & 1 \end{bmatrix},\ B_1 = B_2 = I_2 \otimes \begin{bmatrix} 0 \\ 1 \end{bmatrix},\ B_3 = I_2,$$

$$B_4 = B_5 = I_2 \otimes \begin{bmatrix} 0 & 0 \\ 0 & 1 \\ 1 & 0 \end{bmatrix},\ B_6 = B_7 = I_2 \otimes \begin{bmatrix} 0 & 0 \\ 1 & 0 \\ 0 & 1 \end{bmatrix},$$

$$C_1 = C_2 = I_2 \otimes \begin{bmatrix} 1\ 0 \end{bmatrix},\ C_3 = I_2,\ C_4 = C_5 = C_6 = C_7 = I_2 \otimes \begin{bmatrix} 1\ 0\ 0 \end{bmatrix},$$

$$D_1 = D_2 = I_4,\ D_3 = I_2,\ D_4 = D_5 = D_6 = D_7 = I_6,$$

$$S = I_2 \otimes \begin{bmatrix} 0 & 0 \\ 0 & -1 \end{bmatrix},\ R = I_2 \otimes \begin{bmatrix} 1\ 1 \end{bmatrix}.$$

The communication topology for the multiagent system is shown in Fig. 6.1a, where the nodes 1–7 denote the followers and the nodes 8–10 denote the leaders. In addition, label $\mathcal{F}$ and $\mathcal{L}$ as the follower set and the leader set, namely, $\mathcal{F} = \{1, \cdots, 7\}$, $\mathcal{L} = \{8, 9\}$.

Obviously, Assumptions 6.1–6.3 are satisfied. Calculate the solution pairs (X_i, U_i) as

$$X_1 = X_2 = I_2 \otimes \begin{bmatrix} 1 & 0 \\ 0 & 1 \end{bmatrix},\ X_3 = I_2 \otimes \begin{bmatrix} 1\ 0 \end{bmatrix},\ X_4 = X_5 = X_6 = X_7 = I_2 \otimes \begin{bmatrix} 1 & 0 \\ 0 & 1 \\ 0 & 0 \end{bmatrix},$$

$$U_1 = U_2 = 0_{2\times 4},\ U_3 = I_2 \otimes \begin{bmatrix} 0\ 1 \end{bmatrix},\ U_4 = U_5 = I_2 \otimes \begin{bmatrix} 1 & 2 \\ 0 & 0 \end{bmatrix},\ U_6 = U_7 = I_2 \otimes \begin{bmatrix} 0 & 0 \\ -2 & 2 \end{bmatrix}.$$

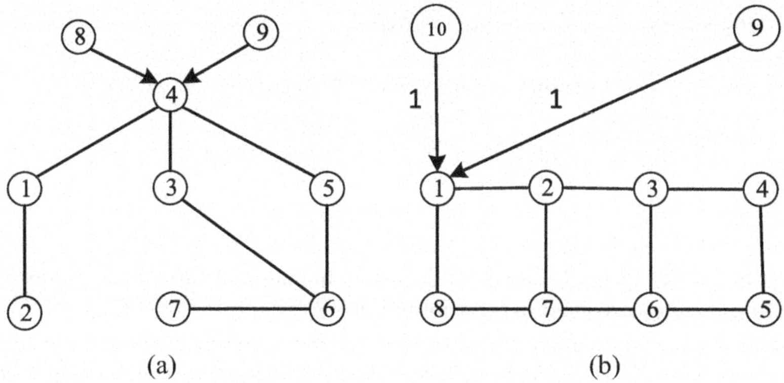

Fig. 6.1 The signed communication topologies $\bar{\mathcal{G}}_s$ for Examples 1–2

Then, calculate the relative matrices as

$$K = 1_2 \otimes [7.0699\ 0; 6.4488\ 0]\,, K_{x_1} = K_{x_2} = I_2 \otimes \begin{bmatrix} -2 & -2 \end{bmatrix},\ K_{x_3} = -I_2,$$

$$K_{x_4} = K_{x_5} = I_2 \otimes \begin{bmatrix} 1 & 2 & -1 \\ -2 & -2 & -1 \end{bmatrix},$$

$$K_{x_6} = K_{x_7} = I_2 \otimes \begin{bmatrix} -2 & -2 & -1 \\ -2 & 2 & -3 \end{bmatrix},$$

$$G_1 = G_2 = I_2 \otimes \mathrm{col}(-9, -20),$$

$$G_3 = -5I_2,\ G_4 = G_5 = I_2 \otimes \mathrm{col}(-14, -58, -33),$$

$$G_6 = G_7 = I_2 \otimes \mathrm{col}(-16, -88, -178).$$

Based on the pole assignment method, select the matrices K_{x_i} and G_i. Select the eigenvalues of the matrices $A_i + B_i K_{x_i}$ as $\{-50 + \mathbf{i}; -50 - \mathbf{i}; -100 + \mathbf{i}; -100 - \mathbf{i}\}$, $\{-50 + \mathbf{i}; -50 - \mathbf{i}; -100 + \mathbf{i}; -100 - \mathbf{i}\}$, $\{-100; -100\}$, $\{-50 + \mathbf{i}; -50 - \mathbf{i}; -50 + \mathbf{i}; -50 - \mathbf{i}; -50; -50\}$, $\{-50 + \mathbf{i}; -50 - \mathbf{i}; -50 + \mathbf{i}; -50 - \mathbf{i}; -50; -50\}$, $\{-50 + \mathbf{i}; -50 - \mathbf{i}; -50 + \mathbf{i}; -50 - \mathbf{i}; -50; -50\}$, $\{-50 + \mathbf{i}; -50 - \mathbf{i}; -50 + \mathbf{i}; -50 - \mathbf{i}; -50; -50\}$, and the eigenvalues of the matrices $A_i + G_i C_i$ as $\{-100; -90; -90; -100\}$, $\{-100; -90; -90; -100\}$, $\{-100; -100\}$, $\{-100; -100; -70; -70; -90; -90\}$, $\{-100; -100; -70; -70; -90; -90\}$, $\{-100; -100; -70; -70; -90; -90\}$, $\{-100; -100; -70; -70; -90; -90\}$, respectively. Give the desired time-varying output formation vectors

$$f_{d_i} = Rf_i = \begin{bmatrix} -1/2 * \sin(t + i * 2\pi/7) \\ 1/2 * \cos(t + i * 2\pi/7) \end{bmatrix},\quad i = 1, 2, 3, 4, 5, 6, 7$$

in which the vectors f_i and f_j are given by

$$f_i = \begin{bmatrix} -1/2\sin(t + i * 2\pi/7) - 1/2\cos(t + i * 2\pi/7) \\ 1/2\cos(t + i * 2\pi/7) \\ -1/2\sin(t + i * 2\pi/7) + 1/2\cos(t + i * 2\pi/7) \\ 1/2\sin(t + i * 2\pi/7) \end{bmatrix}$$

Clearly, the given formation conditions satisfy the formation feasibility condition and the trajectories of the followers will form a heptagonal formation structure.

The initial states involved in the simulation are randomly taken from the interval (0, 1), and the initial adaptive weights and internal dynamic variables are selected as $\hat{c}_i(0) = \eta_i(0) = 0.01$ and $\bar{c}_i = 4.8$, and the other parameters are selected as $c = 6.0533$, $\gamma_i = 1$, $\sigma_i = 0.85$, $\beta_i = 1/\gamma_{i0}$, $\rho_i = 0.01$, $\theta_i = \xi_i = \gamma_{i0}$, $i \in \mathcal{F}$, which are taken in turn from the set [710.0772; 236.6924; 710.0772; 3550.4; 710.0772; 1420.2; 236.6924]. The disturbances in Eq. (6.1) are given by

$$\omega_1 = \text{col}(3.567, 0.9384, 2.4860, 1.6151),$$

$$\omega_2 = \text{col}(1.0426, 1.4657, 1.8584, 3.4739),$$

$$\omega_3 = \text{col}(3.1200, 4.9870),$$

$$\omega_4 = \text{col}(0.6950, 1.8643, 1.8292, 3.2239, 4.3081, 2.8280),$$

$$\omega_5 = \text{col}(0.1000, 3.6563, 3.7331, 4.0263, 2.1948, 0.8927),$$

$$\omega_6 = \text{col}(1.7616, 3.1267, 1.2165, 4.4841, 4.0868, 4.6285),$$

$$\omega_7 = \text{col}(0.1964; 2.6488; 3.5029, 3.6382, 0.9484, 2.5878).$$

Then, the simulation results are shown in Figs. 6.2, 6.3, 6.4, 6.5, 6.6, and 6.7.

Figure 6.2 illustrates the dynamic response of the adaptive weights, from which it can be observed that the adaptive weights eventually converge to the finite steady values. Figure 6.3 illustrates the dynamic trajectory of the internal dynamical variables, from which it can be observed that the internal dynamical variables eventually converge to zero, which argues for the feasibility of the conclusions of Lemma 6.1. Figure 6.4 shows the dynamic trajectory of the compensator error, and it is clear that the compensator error eventually converges to zero, suggesting that the compensator designed in this chapter can successfully estimate the convex combination of the leaders. Figure 6.5 shows the moments when events occur for each follower under the dynamic event-triggering mechanism, and it is clear that each follower will not be triggered sequentially, and thus the triggering mechanism proposed in this chapter will not fall into the Zeno phenomenon. Figure 6.6 shows the dynamic trajectory

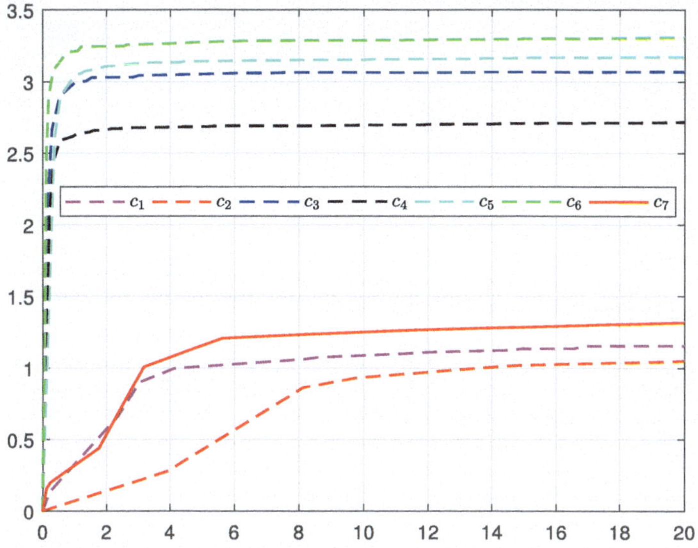

Fig. 6.2 Example 6.1: The trajectory of adaptive coupling weight $\hat{c}_i$, reprinted from Zhang et al. (2024), copyright 2024, with permission from IEEE

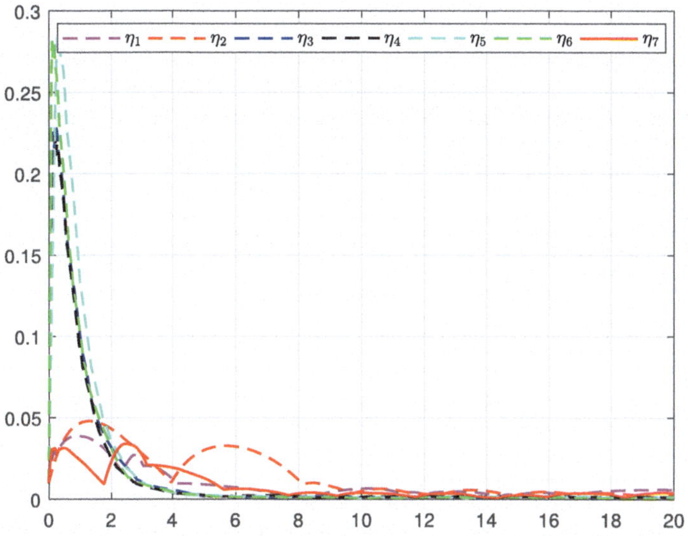

Fig. 6.3 Example 6.1: The trajectory of internal dynamic variable η_i, reprinted from Zhang et al. (2024), copyright 2024, with permission from IEEE

of the formation tracking error, and it is clear that the formation error eventually converges and tends to zero, which indicates that the formation tracking controller designed in this chapter is feasible. In addition, to observe the trajectories of all the agents more intuitively, we plot the out snapshots of 9 agents at different time

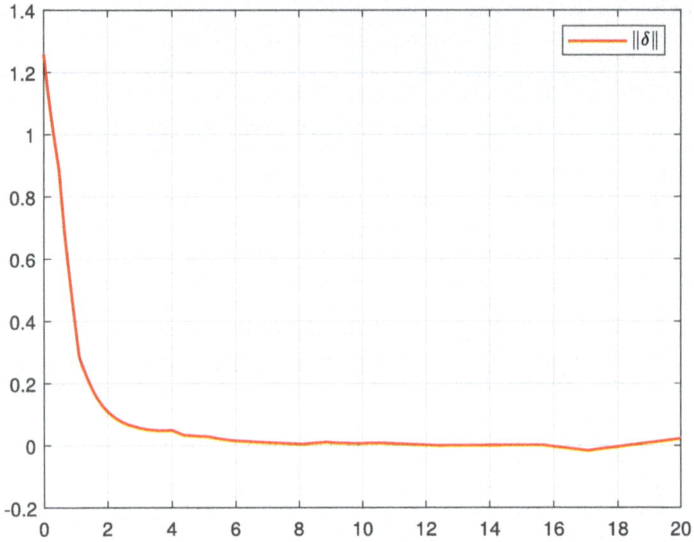

Fig. 6.4 Example 6.1: The time evolution of $\|\delta\|$ for the norm of compensator error, reprinted from Zhang et al. (2024), copyright 2024, with permission from IEEE

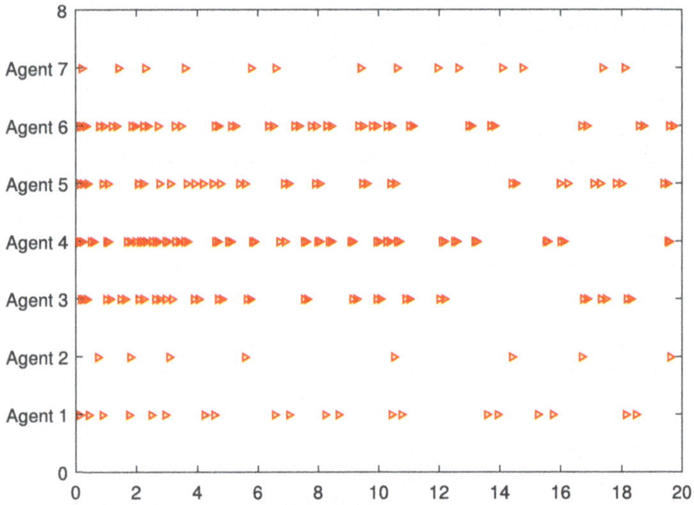

Fig. 6.5 Example 6.1: The triggering instants for all followers, reprinted from Zhang et al. (2024), copyright 2024, with permission from IEEE

instants, from which it can be observed that the follower tracks the leader's signal and eventually forms a heptagonal formation.

The advantage of the dynamic event-triggered mechanism (DETM) designed in this chapter is to expand the triggering threshold by introducing internal dynamic

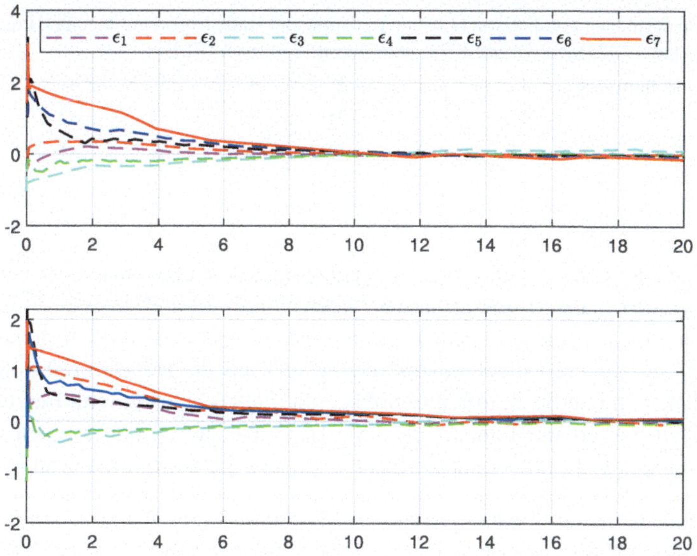

Fig. 6.6 Example 6.1: The time evolution of output formation error ϵ, reprinted from Zhang et al. (2024), copyright 2024, with permission from IEEE

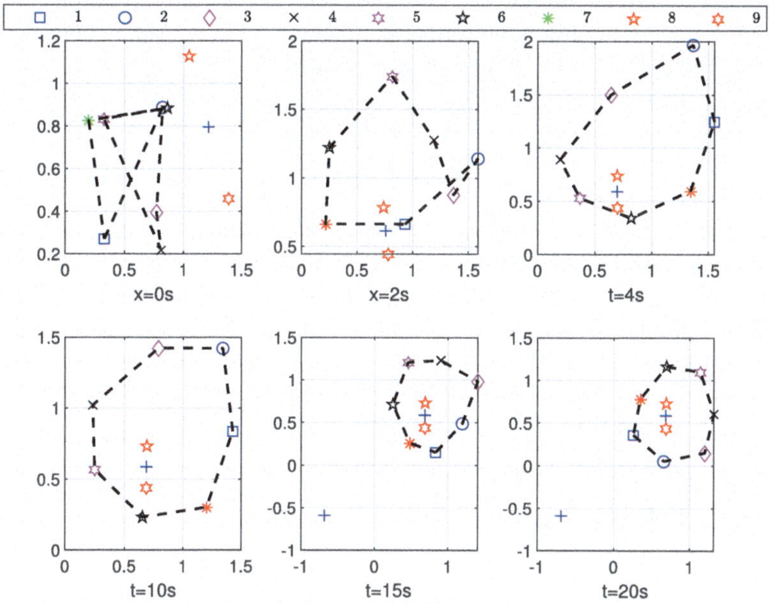

Fig. 6.7 Example 6.1: Out snapshots of the nine agents, reprinted from Zhang et al. (2024), copyright 2024, with permission from IEEE

Table 6.1 Performance comparison between dynamic and static triggering mechanisms, reprinted from Zhang et al. (2024), copyright 2024, with permission from IEEE

Triggering mechanisms	1	2	3	4	5	6	7
DETM	15	10	21	72	32	42	20
SETM	45	39	42	101	61	85	73

variables, which in turn expands the triggering interval of adjacent events, reduces the frequency of information transmission between neighbors, and saves system communication resources. Table 6.1 records the number of triggered for 7 followers under different event-triggered mechanisms (trying to keep the same initial conditions). From Table 6.1, it can be significantly observed that comparing with the mentioned static event-triggered mechanism (SETM) ($\eta_i = 0$), the dynamic event-triggered mechanism proposed in this chapter greatly reduces the communication frequency.

Example 6.2 Consider the heterogeneous MAS consisting of a group of UGVs and UAVs, in which 8 followers ($\mathcal{F} = \{1, \cdots, 8\}$) and 2 leaders ($\mathcal{L} = \{9, 10\}$) need to perform air-ground cooperative formation tasks. The communication way between followers and leaders is shown in Fig. 6.1b. Two leaders ($\mathcal{L} = \{9, 10\}$) and four followers ($\mathcal{F}_1 = \{5, \cdots, 8\}$) are selected as quadrotor UAVs, and the remaining four followers $\mathcal{F}_2 = \{1, \cdots, 4\}$ are selected as quadrotor UGVs. Assume that all UAVs fly at a predefined constant altitude $h = 5m$, and each UAV has no formation controller along the $Z-$axis. Therefore, the formation tracking problem can be considered on XY plane. The dynamics of UAVs can be described by

$$\begin{aligned} \dot{\mathfrak{r}}_i &= \nu_i, \ i \in \mathcal{F}_1 \cup \mathcal{L} \\ \dot{\nu}_i &= \alpha_{\mathfrak{r}i}\mathfrak{r}_i + \alpha_{\nu i}\nu_i + u_i \end{aligned}$$

where $\mathfrak{r}_i \in R^2, \nu_i \in R^2, \alpha_{\mathfrak{r}i} \in R, \alpha_{\nu i} \in R$ express the position, velocity, two damping constants, respectively. The dynamics of UGVs can be described by $\dot{\mathfrak{r}}_j = u_j, \ j \in \mathcal{F}_2$, where $\mathfrak{r}_j \in R^2$ express the position. Based on the outer/inner loop framework in Dong et al. (2017) and the linearization method in Ren and Nathan (2008), the leaders are modeled as $S = I_2 \otimes [0\ 1; -1\ 0]$, $R = I_2 \otimes [1\ 0]$, by selecting $\alpha_{\mathfrak{r}9} = \alpha_{\mathfrak{r}10} = -1, \alpha_{\nu 9} = \alpha_{\nu 10} = 0$, and $u_9 = u_{10} = 0$, the followers $\mathcal{F}_1$ are modeled by $A_i = I_2 \otimes \begin{bmatrix} 0\ 1 \\ 0\ 0 \end{bmatrix}$, $B_i = I_2 \otimes \begin{bmatrix} 0 \\ 1 \end{bmatrix}$, $C_i = I_2 \otimes \begin{bmatrix} 1\ 0 \end{bmatrix}$, $D_i = 0_{4\times 4}$, and the followers $\mathcal{F}_2$ are modeled by $A_j = 0_{2\times 2}$, $B_j = I_2$, $C_j = I_2$, $D_j = 0_{2\times 2}$. The followers $\mathcal{F}_1$ and $\mathcal{F}_2$, respectively, implement regular quadrilateral formation, f_i ($i \in \mathcal{F}_1 \cup \mathcal{F}_2$) are given by

$$f_i = Rf_i = \begin{bmatrix} 3\cos(t + (i-5)\pi/2) \\ -3\sin(t + (i-5)\pi/2) \\ 3\sin(t + (i-5)\pi/2) \\ 3\cos(t + (i-5)\pi/2) \end{bmatrix}.$$

Select $K_{x_i} = I_2 \otimes [-2 \quad -2], i \in \mathcal{F}_1$, $K_{x_j} = -I_2, j \in \mathcal{F}_2$, $K = I_2 \otimes \begin{bmatrix} -1.4142 & 0 \\ 0 & 1.4142 \end{bmatrix}$, $G_i = I_2 \otimes \begin{bmatrix} -9 \\ -20 \end{bmatrix}$, $i \in \mathcal{F}_1$, $G_j = -5I_2$, $j \in \mathcal{F}_2$. The initial states involved in the simulation are randomly taken from the interval (0, 1), other relevant parameters are chosen according to Algorithm 6.1, Theorems 6.1 and 6.2. Then, the simulation results are shown in Figs. 6.8, 6.9, 6.10, and 6.11.

Figure 6.8 depicts the trajectory of the adaptive weight, from which it can be observed that the adaptive weight eventually converges to the stable value. Figure 6.9 depicts the trajectory of the compensator error, from which it can be observed that the compensator error eventually converges to 0. Hence, the designed compensator in Eq. (6.5) is feasible. Figure 6.10 depicts the triggering instants of the different followers, and it is clear that the multiagent system does not exhibit Zeno behavior. Figure 6.11 depicts the dynamic response of the output formation tracking error, and it is clear that the formation tracking error eventually converges to 0, and thus the formation tracking problem is solved with the proposed controller.

6.5 Summary

This chapter investigated the time-varying output formation tracking problem for heterogeneous linear multiagent systems with disturbances. The adaptive distributed dynamic event-triggered compensator was designed for each follower to estimate

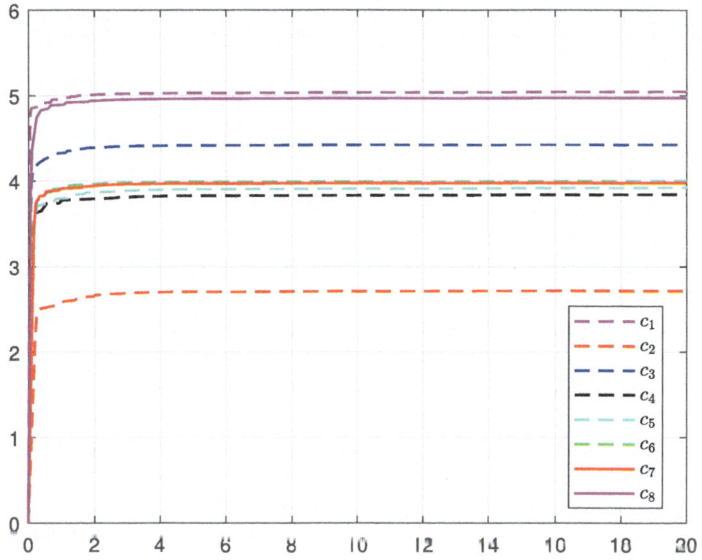

Fig. 6.8 Example 6.2: The trajectory of adaptive weight $\hat{c}_i$, reprinted from Zhang et al. (2024), copyright 2024, with permission from IEEE

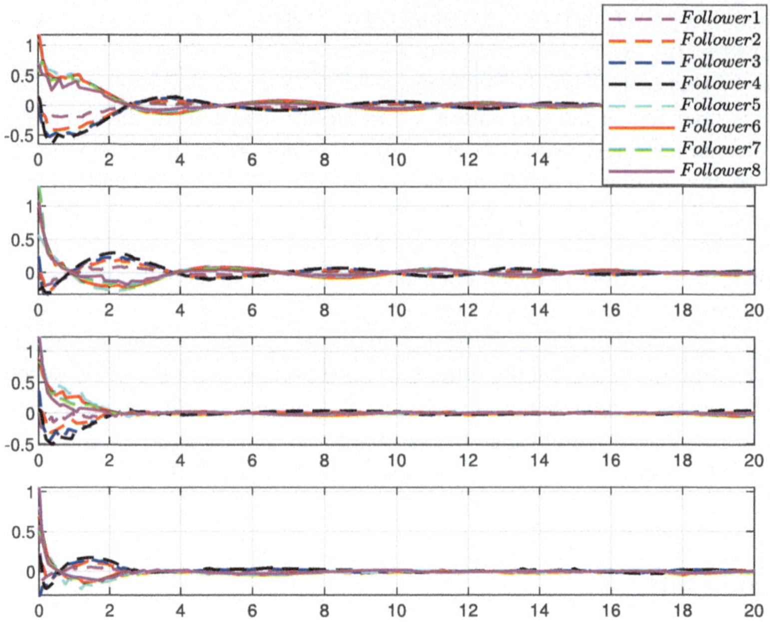

Fig. 6.9 Example 6.2: The trajectory of compensator error δ, reprinted from Zhang et al. (2024), copyright 2024, with permission from IEEE

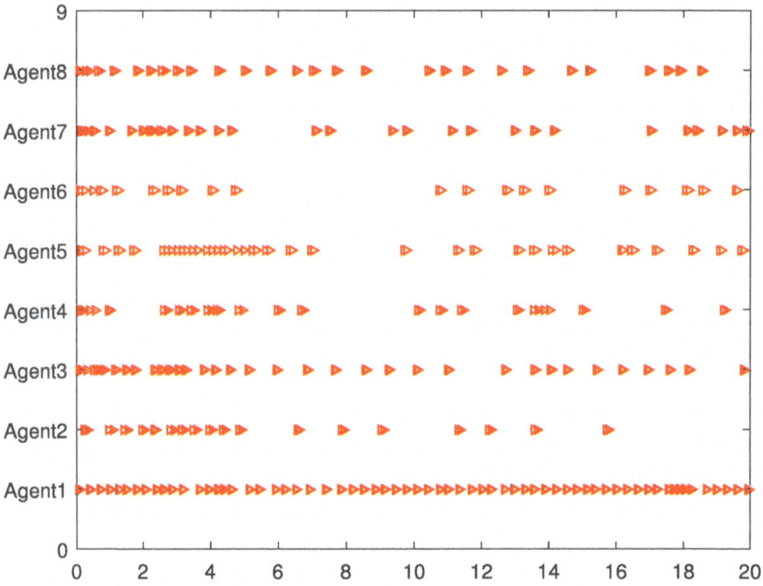

Fig. 6.10 Example 6.2: The triggering instants for all followers, reprinted from Zhang et al. (2024), copyright 2024, with permission from IEEE

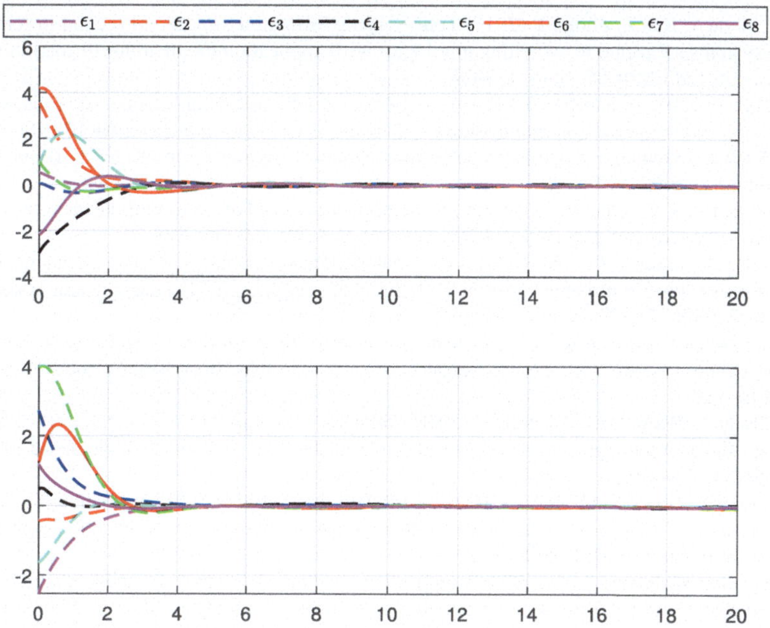

Fig. 6.11 Example 6.2: The time evolution of output formation tracking error ϵ, reprinted from Zhang et al. (2024), copyright 2024, with permission from IEEE

the convex combination formed by all leaders. Theorem 6.1 and simulation results together show that the designed adaptive distributed dynamic event-triggered compensator is feasible, and the designed dynamic event-triggered mechanism does not make multiagent system exhibit Zeno behavior and significantly reduces the communication frequency as compared to the static event-triggered mechanism. Theorem 6.2 and simulation results show that the designed control protocol successfully solves the time-varying output formation tracking problem and realizes present formations.

References

Cheng B, Wang X, Li Z (2019) Event-triggered consensus of homogeneous and heterogeneous multiagent systems with jointly connected switching topologies. IEEE Trans Cybern 49(12):4421–4430

Dimarogonas DV, Frazzoli E, Johansson KH (2012) Distributed event-triggered control for multiagent systems. IEEE Trans Autom Control 57(5):1291–1297

Ding L, Han Q-L, Ge X, Zhang X-M (2018) An overview of recent advances in event-triggered consensus of multiagent systems. IEEE Trans Cybern 48(4):1110–1123

Ding L, Han Q-L, Zhang X-M (2019) Distributed secondary control for active power sharing and frequency regulation in islanded microgrids using an event-triggered communication mechanism. IEEE Trans Ind Inf 15(7):3910–3922

Dong X, Zhou Y, Ren Z, Zhong Y (2017) Time-varying formation tracking for second-order multi-agent systems subjected to switching topologies with application to quadrotor formation flying. IEEE Trans Ind Electr 64(6):5014–5024

Ge X, Han Q (2017) Distributed formation control of networked multi-agent systems using a dynamic event-triggered communication mechanism. IEEE Trans Ind Electr 64(10):8118–8127

Girard A (2015) Dynamic triggering mechanisms for event-triggered control. IEEE Trans Autom Control 60(7):1992–1997

He W, Xu B, Han Q-L, Qian F (2020) Adaptive consensus control of linear multiagent systems with dynamic event-triggered strategies. IEEE Trans Cybern 50(7):2996–3008

Hu W, Yang C, Huang T, Gui W (2020) A distributed dynamic event-triggered control approach to consensus of linear multiagent systems with directed networks. IEEE Trans Cybern 50(2):869–874

Hu J, Bhowmick P, Lanzon A (2020) Distributed adaptive time-varying group formation tracking for multiagent systems with multiple leaders on directed graphs. IEEE Trans Control Netw Syst 7(1):140–150

Hua Y, Dong X, Wang J, Li Q, Zhang R (2018) Time-varying output formation tracking of heterogeneous linear multi-agent systems with multiple leaders and switching topologies. J Franklin Inst 356(1):539–560

Hua Y, Dong X, Hu G, Li Q, Ren Z (2019) Distributed time-varying output formation tracking for heterogeneous linear multiagent systems with a nonautonomous leader of unknown input. IEEE Trans Autom Control 64(10):4292–4299

Huang J, Wang W, Wen C, Li G (2020) Adaptive event-triggered control of nonlinear systems with controller and parameter estimator triggering. IEEE Trans Autom Control 65(1):318–324

Jiang W, Wen G, Peng Z, Huang T, Rahmani A (2019) Fully distributed formation-containment control of heterogeneous linear multiagent systems. IEEE Trans Autom Control 64(9):3889–3896

Li Z, Liu X, Lin P, Ren W (2011) Consensus of linear multi-agent systems with reduced-order observer-based protocols. Syst Control Lett 60(2011):510–516

Li M, Zhang W, Yan C, Hu Z (2022) Observer-based bipartite formation control for mass with external disturbances under event-triggered scheme. IEEE Trans Circuits Syst II: Exp Brief 69(3):1178–1182

Liang H, Zhang L, Sun Y, Huang T (2021) Containment control of semi-markovian multiagent systems with switching topologies. IEEE Trans Syst Man Cybern Syst 51(6):3889–3899

Liang J, Bu X, Cui L, Hou Z (2023) Data-driven bipartite formation for a class of nonlinear mimo multiagent systems. IEEE Trans Neural Netw Learn Syst 34(6):3161–3173

Liu D, Yang G-H (2021) A dynamic event-triggered control approach to leader-following consensus for linear multiagent systems. IEEE Trans Syst Man Cybern Syst 51(10):6271–6279

Liu C, Wu X, Wan X, Lv J (2021) Time-varying output formation tracking of heterogeneous linear multi-agent systems with dynamical controllers. Neurocomputing 441(21):36–43

Liu B, Ni W, Liu RP, Guo YJ, Zhu H (2023) Decentralized privacy-preserving routing of cellular-connected unmanned aerial vehicles for joint goods delivery and sensing. IEEE Trans Intell Transp Syst 24(9):9627–9641

Peng C, Sun H (2020) Switching-like event-triggered control for networked control systems under malicious denial of service attacks. IEEE Trans Autom Control 65(9):3943–3949

Ren W, Nathan S (2008) Distributed coordination architecture for multi-robot formation control. Robot Autonom Syst 56(4):324–333

Wang R, Dong X, Li Q, Ren Z (2019) Distributed time-varying formation control for multiagent systems with directed topology using an adaptive output-feedback approach. IEEE Trans Ind Inf 15(8):4676–4685

Xie X, Yue D, Park JH (2020) Observer-based state estimation of discrete-time fuzzy systems based on a joint switching mechanism for adjacent instants. IEEE Trans Cybern 50(8):3545–3555

Yan C, Zhang W, Su H, Li X (2022) Adaptive bipartite time-varying output formation control for multiagent systems on signed directed graphs. IEEE Trans Cybern 52(9):8987–9000

Yang Y, Modares H, Vamvoudakis KG, Yin Y, Wunsch DC (2020) Dynamic intermittent feedback design for h_∞ containment control on a directed graph. IEEE Trans Cybern 50(8):3752–3765

Zhang J, Zhang H, Sun S (2024) Adaptive dynamic event-triggered bipartite time-varying output formation tracking problem of heterogeneous multiagent systems. IEEE Trans Syst Man Cybern Syst 54(1):12–22

Zhang J, Zhang H, Liang Y, Song W (2023b) Adaptive bipartite output tracking consensus in switching networks of heterogeneous linear multiagent systems based on edge events. IEEE Trans Neural Netw Learn Syst 34(1):79–89

Zhang J, Zhang H, Ming Z, Mu Y (2023a) Adaptive event-triggered time-varying output bipartite formation containment of multiagent systems under directed graphs. IEEE Trans Neural Netw Learn Syst 34(11):8909–8922

Zhang J, Zhang H, Sun S, Cai Y (2022) Adaptive time-varying formation tracking control for multiagent systems with nonzero leader input by intermittent communications. IEEE Trans Cybern 1–10. https://doi.org/10.1109/TCYB.2022.3165212

Zhao Y, Xian C, Wen G, Huang P, Ren W (2022) Design of distributed event-triggered average tracking algorithms for homogeneous and heterogeneous multiagent systems. IEEE Trans Autom Control 67(3):1269–1284

Zuo S, Song Y, Lewis FL, Davoudi A (2019) Time-varying output formation containment of general linear homogeneous and heterogeneous multiagent systems. IEEE Trans Control Netw Syst 6(2):537–548

Zuo S, Song Y, Lewis FL, Davoudi A (2019) Time-varying output formation containment of general linear homogeneous and heterogeneous multiagent systems. IEEE Trans Control Netw Syst 6(2):537–548

Chapter 7
Formation Containment Control of Multiagent Systems Under Directed Topology

7.1 Introduction

Chapters 2 to 5 investigate the leaderless consensus problem, the leader-follower consensus problem, the output regulation problem, the containment problem, and the formation tracking problem, respectively. Cooperative control problems for multiagent systems also include formation containment problem. Generally speaking, formation problems are divided into two categories: time-invariant and time-varying formation problems in Cheng et al. (2021), Wang et al. (2017). In some practical scenarios, such as source detection, obstacle avoidance, etc., the predetermined formation needs to change over time according to the task requirements or the external environment. Therefore, it is more useful to study the time-varying formation containment problem than the time-invariant formation containment problem.

Some research results have been achieved on the formation problem of multiagent systems. For example, the authors in Hua et al. (2018), Zuo et al. (2019), Dong and Hu (2017) have studied the formation problem of homogeneous or heterogeneous linear multiagent systems under directed topology graphs, where the leader has zero control input. In these works, the designed control protocols rely on the global information of the communication topology graph and require continuous communication between agents, so the proposed approach is not applicable to multiagent systems with large networks and bandwidth constraints. The authors in Jiang et al. (2021), Hu et al. (2020) adopt the adaptive mechanisms to investigate the formation containment problem of linear multiagent systems under directed topology graphs, and the proposed method does not rely on any global information, but the agent still needs to communicate with each other continuously. Up to now, the distributed adaptive event-triggered control law for solving the formation problem of multiagent systems is still limited to the case of undirected topology graphs. Therefore, solving the formation containment problem of linear multiagent systems with directed topology using distributed adaptive event-triggered control laws is still an open problem that requires further research.

J. Zhang et al., *Adaptive Collaborative Control of Multiagent Systems by Event-Triggered Mechanisms*, Intelligent Control and Learning Systems 14,
https://doi.org/10.1007/978-981-97-7866-9_7

To fill the above gap, this chapter investigates the formation containment problem for heterogeneous linear multiagent systems under directed topology graphs based on adaptive event-triggered control algorithms, where the proposed method does not rely on the global information of the communication topology graph. The main challenges are that the Laplace matrix under a directed communication topology graph is asymmetric, and the adaptive event-triggered control laws proposed in Qian et al. (2021), Cheng et al. (2021), Zhang et al. (2020), Cheng and Li (2019), Zhang et al. (2022) cannot be directly used to solve the formation containment problems considered in this chapter, and they are limited to undirected topology graphs. How to design an appropriate event-triggered mechanism to determine the sampling time and whose construction does not depend on the global information of the communication topology graph is challenging.

7.2 Problem Formulation

Consider a heterogeneous linear multiagent system consisting of one virtual leader (labeled 0), N formation leaders, and M followers. The dynamics of the virtual leader is given as

$$\dot{x}_0(t) = A_0 x_0(t), \;\; y_0(t) = C_0 x_0(t), \tag{7.1}$$

the dynamics of the formation leaders and followers are modeled by

$$\begin{aligned} \dot{x}_i(t) &= A_i x_i(t) + B_i u_i(t), \\ y_i(t) &= C_i x_i(t), \;\; i \in I[1, N+M], \end{aligned} \tag{7.2}$$

where $x_0(t) \in R^n$, $y_0(t) \in R^s$, respectively, denote the virtual leader state and output, $x_i(t) \in R^{n_i}$, $u_i(t) \in R^{m_i}$, $y_i(t) \in R^s$, respectively, denote the ith agent state, input and output, and A_0, C_0, A_i, B_i, C_i are the system matrices with compatible dimensions.

Formation containment problem: Consider the multiagent systems composed of Eqs. (7.1)and (7.2) under directed topology graphs. Design the distributed control law such that the closed-loop system is asymptotically stable for any given bounded initial state. Moreover, the following conditions hold:

$$\lim_{t\to\infty} \|y_i(t) - y_0(t) - C_0 h_i(t)\| = 0, \; i \in I[1, N], \tag{7.3a}$$

$$\lim_{t\to\infty} \inf_{y(t)\in Co(y_{\mathfrak{L}}(t))} \|y_i(t) - y(t)\| = 0, \; i \in I[N+1, N+M], \tag{7.3b}$$

where $Co(y_{\mathfrak{L}}(t))$ denotes the convex hull of the formation leaders, $y_{\mathfrak{L}} = \mathrm{col}(y_1, \cdots, y_N)$, $h_i(t)$ denotes the ith ($i \in I[1, N]$) leader's ideal formation vector.

Then, the formation containment problem of the multiagent system is said to be solved.

In order to realize the formation containment problem for multiagent systems, the following assumptions are given.

Assumption 7.1 The system matrices (A_i, B_i, C_i) are stabilizable and detectable, (A_0, C_0) is detectable, and $\text{rank}(B_i) = m_i$, $i \in I[1, N]$.

Assumption 7.2 For the directed graph $\tilde{\mathcal{G}}$, there exists at least one informed leader who has a path to any uninformed leader, and at least one leader who has a path to any follower.

Assumption 7.3 The pairs (Π_i, Γ_i) are the solutions of the following regulator equations:

$$\begin{aligned} \Pi_i A_0 &= A_i \Pi_i + B_i \Gamma_i \\ C_0 &= C_i \Pi_i, \ i \in I[1, N]. \end{aligned} \tag{7.4}$$

7.3 Main Results

In this section, the subsystems that make up the multiagent system are categorized according to their tasks into the virtual leader (preplanning the overall trajectory), the formation leaders (executing of formation tasks), and the followers (entering the desired area). The virtual leader is the center of the formation. Firstly, the observer is designed for the formation leader to estimate the state of the virtual leader. Secondly, the control law is designed for the formation leader to form the desired formation based on the given observer. Finally, the design of the containment control law for each follower makes it possible to converge to the convex hull formed by the formation leaders.

7.3.1 Formation Control Law Design for the Formation Leaders

In order for the formation leader to form the desired formation and keep track of the virtual leader, the formation leader needs to obtain the state information of the virtual leader. Design the observer for each formation leader, where v_i denotes the observer state. Define the observer error $e_{v_i} = v_i - x_0$, $i \in \mathfrak{L}$. Formation leaders are categorized into informed and uninformed formation leaders based on whether or not they have direct access to information about the virtual leader. In the following, observers are designed for these two types of formation leaders separately to estimate the state of the virtual leader.

The following observer is designed for the informed formation leader:

$$\dot{v}_i = A_0 v_i + S(C_0 v_i - y_0),\ i \in I[1, O], \tag{7.5}$$

where v_i denotes the observer state, S is the feedback gain matrix, and S is chosen such that the matrix $A_0 + SC_0$ is Hurwitz.

The following observer is designed for the uninformed formation leader:

$$\dot{v}_i = A_0 v_i + c_i K \tilde{v}_i(t_k^i),\ \ c_i = \alpha a_i + \beta b_i(t_k^i),\ \ i \in I[O+1, N], \tag{7.6}$$

where $\tilde{v}_i = \sum_{j=1}^{N} a_{ij}(v_i - v_j)$, c_i denotes the adaptive weight, K is the feedback gain matrix to be designed, α, β are the positive normal number, $a_i \in R$ and $b_i \in R$ satisfy

$$\dot{a}_i = \gamma \|K\tilde{v}_i(t_k^i)\|^2,\ t \in [t_k^i, t_{k+1}^i), \tag{7.7a}$$

$$b_i = \tilde{v}_i^T Q \tilde{v}_i, \tag{7.7b}$$

$\gamma > 0\ (\in R)$ is the given constant, and $Q > 0$ is the positive definite matrix to be designed.

Remark 7.1 Let $\tilde{v} = \mathrm{col}(\tilde{v}_{O+1}, \cdots, \tilde{v}_N)$. $e_{V_1} = \mathrm{col}(e_{v_{O+1}}, \cdots, e_{v_N})$, $e_{V_2} = \mathrm{col}(e_{v_1}, \cdots, e_{v_O})$. Define the observer error $e_V = \mathrm{col}(e_{V_1}, e_{V_2})$. Therefore, one can obtain that $\tilde{v}$ satisfies $\tilde{v} = (\tilde{\mathcal{L}}_1 \otimes I)e_{V_1} + (\tilde{\mathcal{L}}_2 \otimes I)e_{V_2}$. On the other hand, according to Eq. (7.5), one can calculate the derivative of e_{v_i} as $\dot{e}_{v_i} = (A_0 + SC_0)e_{v_i}$, $i \in I[1, O]$. Since $A + SC_0$ is Hurwitz, one can obtain that $\lim_{t\to\infty} e_{v_i} = 0$, $i \in I[1, O]$, i.e., $\lim_{t\to\infty} e_{V_2} = 0$. Since $\tilde{\mathcal{L}}_1$ is nonsingular based on Lemma 1.4. Therefore, $\lim_{t\to\infty} \tilde{v} = 0$ can be obtained if and only if $\lim_{t\to\infty} e_{V_1} = 0$.

Define the measurement error as $e_{\tilde{v}_i} = \tilde{v}_i(t_k^i) - \tilde{v}_i$, $t \in [t_k^i, t_{k+1}^i)$. The triggering instant t_k^i depends on the following event-triggered mechanism:

$$t_{k+1}^i = \{t > t_k^i | h_i^1(t) \geq 0 \text{ or } h_i^2(t) \geq 0\}, \tag{7.8}$$

where $h_i^1(t)$ and $h_i^2(t)$ denote the triggering functions. The triggering functions are designed as

$$h_i^1 = d_i^2 \|K\tilde{v}_i(t_k^i)\|^2 - \theta\|K\tilde{v}_i\|^2 - \mu_{h^1},\ \ h_i^2 = \bar{c}_i^2 \|K e_{\tilde{v}_i}\|^2 - \vartheta\|K\tilde{v}_i\|^2 - \mu_{h^2}, \tag{7.9}$$

where d_i and $\bar{c}_i$ satisfy

$$d_i(t) = b_i(t) - b_i(t_k^i),\quad t \in [t_k^i, t_{k+1}^i), \tag{7.10a}$$

$$\bar{c}_i(t) = \alpha a_i(t) + \beta b_i(t), \tag{7.10b}$$

μ_{h^1} and μ_{h^2} belong to $\mathbb{L}_1[0, +\infty)$ space.

Let $V_1 = \text{col}(v_{O+1}, \cdots, v_N)$, $V_2 = \text{col}(v_1, \cdots, v_O)$. Then, calculate the derivative of V_1 and V_2

$$\dot{V}_2 = (I \otimes A_0)V_2 + (I \otimes SC_0)e_{v_2}, \tag{7.11a}$$

$$\dot{V}_1 = (I \otimes A_0)V_1 + (\mathcal{C} \otimes K)(\tilde{v} + e_{v_1}). \tag{7.11b}$$

Since $\tilde{v}_i = \sum_{j=1}^{N} a_{ij}(v_i - v_j)$, we can obtain the global form as

$$\tilde{v} = (\tilde{\mathcal{L}}_1 \otimes I)V_1 + (\tilde{\mathcal{L}}_2 \otimes I)V_2.$$

In addition, it follows from Eq. (7.11) that the derivative of $\tilde{v}$ satisfies

$$\dot{\tilde{v}} = (I \otimes A_0)\tilde{v} + (\tilde{\mathcal{L}}_1\mathcal{C} \otimes K)(\tilde{v} + e_{v_1}) + (\tilde{\mathcal{L}}_2 \otimes SC_0)e_{v_2}. \tag{7.12}$$

The following lemma ensures that the observer error converges to zero, i.e., $\lim_{t\to\infty} e_{v_i} = 0$ and $i \in \mathfrak{L}$.

Lemma 7.1 *Under the condition of Assumptions 7.1–7.3, if the matrix S is given in Eq. (7.5) with the matrix $A_0 + SC_0$ being Hurwitz, the feedback matrix $K = -Q$, and $Q > 0$ is the solution of the following equation:*

$$QA_0 + A_0^T Q - \frac{1}{2}Q^2 + P = 0, \tag{7.13}$$

where $P > 0$ is the given positive definite matrix, and the initial weight $a_i(0) > \frac{\sqrt{\vartheta}}{\alpha}$ is chosen. Then, one obtains that the observer error e_{v_i} converges to 0, i.e., $\lim_{t\to\infty} e_{v_i} = 0$, $i \in \mathfrak{L}$, and the adaptive weights c_i ($i \in I[O+1, N]$) converge to finite values. Moreover, the multiagent system does not exhibit Zeno behavior.

Proof It follows from Eqs. (7.1) and (7.5), one obtains

$$\dot{e}_{v_i} = (A_0 + SC_0)e_{v_i}, \ i \in I[1, O]. \tag{7.14}$$

It follows from Eq. (7.14) that $\lim_{t\to\infty} e_{v_i} = 0, \ i \in I[1, O]$ since $A_0 + SC_0$ is Hurwitz. Then, we prove that $\lim_{t\to\infty} e_{v_i} = 0, \ i \in I[O+1, N]$. According to Remark 7.1, we obtain that $\lim_{t\to\infty} e_{v_i} = 0, \ i \in I[O+1, N]$ if and only if $\lim_{t\to\infty} \tilde{v} = 0$. Thus, we only need to prove that $\lim_{t\to\infty} \tilde{v} = 0$.

Consider the following Lyapunov function:

$$V = \sum_{i\in I[O+1,N]} \bar{v}_i(\alpha a_i b_i + \frac{1}{2}\beta b_i^2) + \sum_{i\in I[O+1,N]} \frac{\alpha^2}{2\beta}\bar{v}_i(a_i - \bar{a}) \tag{7.15}$$

where $\bar{v}_i$ are given in Lemma 4.1, and $\bar{a} \in R$ is the positive constant to be given. Calculate the derivative of V along Eq. (7.7),

$$\begin{aligned}\dot{V} =& \sum_{i\in I[O+1,N]} \bar{v}_i \bar{c}_i \dot{b}_i + \frac{\alpha}{\beta} \sum_{i\in I[O+1,N]} \bar{v}_i(\beta b_i + \alpha a_i - \alpha\bar{a})\dot{a}_i \\ =& \frac{\alpha}{\beta} \sum_{i\in I[O+1,N]} \bar{v}_i(\beta b_i + \alpha(a_i - \bar{a}))\gamma \|K\tilde{v}_i(t_k^i)\|^2 \\ &+ 2 \sum_{i\in I[O+1,N]} \bar{v}_i \bar{c}_i \tilde{v}_i^T Q \dot{\tilde{v}}_i \\ =& \tilde{v}^T \big(\bar{\Xi}\bar{\mathcal{C}} \otimes (A_0^T Q + Q A_0)\big)\tilde{v} - 2\tilde{v}^T(\bar{\mathcal{C}}\bar{\Xi}\mathcal{L}_1\mathcal{C} \otimes K^T K)(\tilde{v} + e_{V_1}) \\ &- 2\tilde{v}^T(\bar{\mathcal{C}}\bar{\Xi}\mathcal{L}_2 \otimes K S C_0)e_{V_2} \\ &+ \frac{\alpha\gamma}{\beta} \sum_{i\in I[O+1,N]} \bar{v}_i(\bar{c}_i - \alpha\bar{a})\|K\tilde{v}_i(t_k^i)\|^2, \end{aligned} \tag{7.16}$$

where $\bar{\Xi}$ is given in Lemma 1.3. Then, consider the following formula:

$$\begin{aligned}&- 2\tilde{v}^T(\bar{\mathcal{C}}\Xi\mathcal{L}_1\mathcal{C} \otimes K^T K)(\tilde{v} + e_{V_1}) \\ &= -2\tilde{v}^T(\bar{\mathcal{C}}\Xi\mathcal{L}_1\bar{\mathcal{C}} \otimes K^T K)\tilde{v} - 2\tilde{v}^T(\bar{\mathcal{C}}\Xi\mathcal{L}_1\bar{\mathcal{C}} \otimes K^T K)e_{V_1} \\ &+ 2\tilde{v}^T(\bar{\mathcal{C}}\Xi\mathcal{L}_1(\bar{\mathcal{C}} - \mathcal{C}) \otimes K^T K)(\tilde{v} + e_{V_1}). \end{aligned} \tag{7.17}$$

According to Lemma 1.3, we can obtain the following formula:

$$\begin{aligned}&- 2\tilde{v}^T(\bar{\mathcal{C}}\Xi\mathcal{L}_1\bar{\mathcal{C}} \otimes K^T K)\tilde{v} \\ &= -\tilde{v}^T(\bar{\mathcal{C}} \otimes K^T)\big((\Xi\mathcal{L}_1 + \mathcal{L}_1^T\Xi) \otimes I\big)(\bar{\mathcal{C}} \otimes K)\tilde{v} \\ &\le - \lambda_{min}(\Xi\mathcal{L}_1 + \mathcal{L}_1^T\Xi)\tilde{v}^T(\bar{\mathcal{C}}^2 \otimes K^T K)\tilde{v}, \end{aligned} \tag{7.18}$$

where $\lambda_{min}(\Xi\mathcal{L}_1 + \mathcal{L}_1^T\Xi)$ denotes the minimum eigenvalue of the matrix $\Xi\mathcal{L}_1 + \mathcal{L}_1^T\Xi$.

It follows from Eqs. (7.6), (7.10), triggering mechanism in Eq. (7.8), and Lemma 1.6 that the following inequality can be obtained for any $\rho_1,\ \rho_2 > 0\ (\in R)$:

$$\begin{aligned}&- 2\tilde{v}^T(\bar{\mathcal{C}}\Xi\mathcal{L}_1\bar{\mathcal{C}} \otimes K^T K)e_{V_1} \\ \le& \rho_1\tilde{\xi}^T(\bar{\mathcal{C}}\Xi\mathcal{L}_1\mathcal{L}_1^T\Xi\bar{\mathcal{C}} \otimes K^T K)\tilde{\xi} + \vartheta/\rho_1\tilde{\xi}^T(I \otimes K^T K)\tilde{\xi} \\ &+ (N - O)\mu_{h^2}/\rho_1 \end{aligned} \tag{7.19}$$

and

$$\begin{aligned}&2\tilde{v}^T(\bar{\mathcal{C}}\Xi\mathcal{L}_1(\bar{\mathcal{C}} - \mathcal{C}) \otimes K^T K)(\tilde{v} + e_{V_1}) \\ \le& \rho_2\tilde{v}^T(\bar{\mathcal{C}}\Xi\mathcal{L}_1\mathcal{L}_1^T\Xi\bar{\mathcal{C}} \otimes K^T K)\tilde{v} \\ &+ \frac{1}{\rho_2}(\tilde{v} + e_{V_1})^T((\bar{\mathcal{C}} - \mathcal{C})^2 \otimes K^T K)(\tilde{v} + e_{V_1}) \\ \le& \rho_2\lambda_{max}(\Xi\mathcal{L}_1\mathcal{L}_1^T\Xi)\tilde{v}^T(\bar{\mathcal{C}}^2 \otimes K^T K)\tilde{v} \end{aligned}$$

$$
\begin{aligned}
&+\frac{\beta^2}{\rho_2}\sum_{i\in I[O+1,N]} d_i^2\|Kv_i(t_k^i)\|^2\\
\leq&\rho_2\lambda_{max}(\Xi\mathcal{L}_1\mathcal{L}_1^T\Xi)\tilde{v}^T(\bar{\mathcal{C}}^2\otimes K^TK)\tilde{v}\\
&+\frac{\beta^2\theta}{\rho_2}\tilde{v}^T(I\otimes K^TK)\tilde{v}+\beta^2(N-O)\mu_{h^1}/\rho_2,
\end{aligned}
\tag{7.20}
$$

where $\lambda_{max}(\Xi\mathcal{L}_1\mathcal{L}_1^T\Xi)$ denotes the maximum eigenvalue of the matrix $\Xi\mathcal{L}_1\mathcal{L}_1^T\Xi$. Substituting Eqs. (7.18)–(7.20) into Eq. (7.17) yields

$$
\begin{aligned}
&-2\tilde{v}^T(\bar{\mathcal{C}}\Xi\mathcal{L}_1\mathcal{C}\otimes K^TK)(\tilde{v}+e_{V_1})\\
\leq&-\rho_3\tilde{v}^T(\bar{\mathcal{C}}^2\otimes K^TK)\tilde{v}+(\frac{\vartheta}{\rho_1}+\frac{\beta^2\theta}{\rho_2})\tilde{v}^T(I\otimes K^TK)\tilde{v}\\
&+(N-O)\frac{\mu_{h^2}}{\rho_1}+\beta^2(N-O)\frac{\mu_{h^1}}{\rho_2}
\end{aligned}
\tag{7.21}
$$

where $\rho_3=\lambda_{min}(\Xi\mathcal{L}_1+\mathcal{L}_1^T\Xi)-\rho_1\lambda_{max}(\Xi\mathcal{L}_1\mathcal{L}_1^T\Xi)-\rho_2\lambda_{max}(\Xi\mathcal{L}_1\mathcal{L}_1^T\Xi)>0$ can be ensured by selecting appropriate parameters ρ_1 and ρ_2.

Using Lemma 3 in Li et al. (2015), Eqs. (7.7) and (7.10), for $\varpi>0\ (\in R)$, one can obtain

$$
\begin{aligned}
&\frac{\alpha\gamma}{\beta}\sum_{i\in I[O+1,N]} \nu_i(\bar{c}_i-\alpha\bar{a})\|K\tilde{v}_i(t_k^i)\|^2\\
\leq&\frac{\alpha\gamma}{\beta}\sum_{i\in I[O+1,N]} \nu_i\Big((2\bar{c}_i-\frac{\alpha\bar{a}\varpi}{1+\varpi}+\frac{2\bar{c}_i\vartheta}{\bar{c}_i^2}+\frac{\alpha\bar{a}\varpi}{\bar{c}_i^2})\|Kv_i(t)\|^2\\
&+\frac{(2\bar{c}_i+\alpha\bar{a}\varpi)\mu_{h^2}}{\bar{c}_i^2}\Big)\\
\leq&\tilde{v}^T\Big((\frac{2\alpha\gamma}{\beta}\Xi\bar{\mathcal{C}}-(\frac{\alpha^2\gamma\bar{a}\varpi}{\beta(1+\varpi)}-\frac{2\gamma\vartheta}{\beta a_m}-\frac{\gamma\bar{a}\vartheta}{\beta a_m^2})\Xi)\otimes K^TK\Big)\tilde{v}\\
&+\rho_4\mu_{h^2}
\end{aligned}
\tag{7.22}
$$

where $\rho_4=\frac{\gamma}{\beta}\sum_{i\in I[O+1,N]}\nu_i(\frac{2}{a_i(0)}+\frac{\bar{a}\varpi}{a_i^2(0)})$, $a_m=\min_{i\in\mathfrak{F}}\{a_i(0)\}$ and $\varpi\in(0,\frac{a_m^2\alpha^2}{\vartheta}-1)$.

Substituting Eqs. (7.21) and (7.21) into (7.20) yields

$$
\begin{aligned}
\dot{V}\leq&\tilde{v}^T(\bar{\Xi}\bar{\mathcal{C}}\otimes(A_0^TQ+QA_0))\tilde{v}+\tilde{v}^T((-\rho_3'\bar{\mathcal{C}}^2+(\frac{\vartheta}{\rho_1'}+\frac{\beta^2\theta}{\rho_2'})I\\
&+\frac{2\alpha\gamma}{\beta}\bar{\Xi}\bar{\mathcal{C}}-(\frac{\alpha^2\gamma\bar{a}\varpi}{\beta(1+\varpi)}-\frac{2\gamma\vartheta}{\beta a_m}-\frac{\gamma\bar{a}\vartheta}{\beta a_m^2})\bar{\Xi})\otimes K^TK)\tilde{v}\\
&+\frac{\beta^2M\mu_{h^1}}{\rho_2}+(\frac{M}{\rho_1}+\rho_4)\mu_{h^2}-2\tilde{v}^T(\bar{\mathcal{C}}\bar{\Xi}\mathcal{L}_2\otimes KSC_0)e_{V_2}.
\end{aligned}
\tag{7.23}
$$

It follows from Lemma 1.6 that

$$\begin{aligned}
&-2\tilde{v}^T(\bar{C}\Xi\mathcal{L}_2 \otimes KSC_0)e_{V_2} \\
\leq &2e_{V_2}^T(\mathcal{L}_2^T\Xi\bar{C}\mathcal{L}_2 \otimes C_0^TS^TK^TK^{-1}SC_0)e_{V_2} \\
&+\frac{1}{2}\tilde{v}^T(\Xi\bar{C} \otimes K^TK)\tilde{v}. \qquad (7.24)
\end{aligned}$$

Since $a_m \geq a_i(0) > \frac{\sqrt{\vartheta}}{\alpha}$, we select the parameter ϖ such that $\varpi \in (0, \frac{a_m^2\alpha^2}{\vartheta} - 1)$. In addition, $\rho_3 > 0$ could be ensured by selecting the appropriate parameters ρ_1 and ρ_2, and $\frac{\alpha^2\gamma\bar{a}\varpi}{\beta(1+\varpi)} - \frac{2\gamma\vartheta}{\beta a_m} - \frac{\gamma\bar{a}\vartheta}{\beta a_m^2}$ could be sufficiently large by selecting $\bar{a}$. Thus, based on *Lemma 4* in Li et al. (2015), one obtains

$$\begin{aligned}
&-\rho_3\bar{C}^2 + (\frac{\vartheta}{\rho_1} + \frac{\beta^2\theta}{\rho_2})I + \frac{2\alpha\gamma}{\beta}\Xi\bar{C} - (\frac{\alpha^2\gamma\bar{a}\varpi}{\beta(1+\varpi)} - \frac{2\gamma\vartheta}{\beta a_m} - \frac{\gamma\bar{a}\vartheta}{\beta a_m^2})\Xi \\
&\leq -\Xi\bar{C}. \qquad (7.25)
\end{aligned}$$

Substituting Eqs. (7.24) and (7.25) into (7.23) yields

$$\begin{aligned}
\dot{V} \leq &\tilde{v}^T(\Xi\bar{C} \otimes (A_0^TQ + QA_0) - \frac{1}{2}\Xi\bar{C} \otimes K^TK)\tilde{v} + \frac{\beta^2M\mu_{h^1}}{\rho_2} \\
&+ (\frac{M}{\rho_1} + \rho_4)\mu_{h^2} + 2e_{V_2}^T(\mathcal{L}_2^T\Xi\bar{C}\mathcal{L}_2 \otimes C_0^TS^TK^TK^{-1}SC_0)e_{V_2} \\
\leq &-\tilde{v}^T(\Xi\bar{C} \otimes P)\tilde{v} + \frac{\beta^2M\mu_{h^1}}{\rho_2} + (\frac{M}{\rho_1} + \rho_4)\mu_{h^2} \\
&+ 2e_{V_2}^T(\mathcal{L}_2^T\Xi\bar{C}\mathcal{L}_2 \otimes C_0^TS^TK^TK^{-1}SC_0)e_{V_2}. \qquad (7.26)
\end{aligned}$$

Integrating both sides of Eq. (7.26) yields

$$\begin{aligned}
V \leq &V(0) + \int_0^t \Big(\frac{\beta^2M\mu_{h^1}(\tau)}{\rho_2} + (\frac{M}{\rho_1} + \rho_4)\mu_{h^2}(\tau) \\
&+ 2e_{V_2}^T(\mathcal{L}_2^T\Xi\bar{C}\mathcal{L}_2 \otimes C_0^TS^TK^TK^{-1}SC_0)e_{V_2}\Big)d\tau. \qquad (7.27)
\end{aligned}$$

Since $\mu_{h^i}(i = \{1, 2\})$ belong to $\mathbb{L}_1[0, +\infty)$, $\lim_{t\to\infty} e_{V_2} = 0$, we obtain that V is bounded from Eq. (7.27). Based on the definition of V, we conclude that $\bar{c}_i$ and $\tilde{\xi}_i$ are bounded for $i \in I[O + 1, N]$.

On the other hand, it follows from Eq. (7.26) that

$$\begin{aligned}
&\int_0^t \tilde{v}^T(\Xi\bar{C} \otimes P)\tilde{v}d\tau \\
\leq &-V(t) + V(0) + \int_0^t (\frac{\beta^2M\mu_{h^1}(\tau)}{\rho_2} + (\frac{M}{\rho_1} + \rho_4)\mu_{h^2}(\tau)
\end{aligned}$$

$$+ 2e_{V_2}^T(\mathcal{L}_2^T \Xi \bar{C} \mathcal{L}_2 \otimes C_0^T S^T K^T K^{-1} S C_0) e_{V_2}) d\tau. \quad (7.28)$$

Therefore, one can obtain that $\int_0^t \tilde{v}^T(\Xi\bar{C} \otimes P)\tilde{v} d\tau$ is bounded. It is not difficult to find that the derivative of $\tilde{v}^T(\Xi\bar{C} \otimes P)\tilde{v}$ is uniformly continuous. Moreover, it follows from Generalized Barbalat's lemma (Huang et al. 2020) that $\lim_{t\to\infty} \tilde{\xi}^T(\Xi\bar{C} \otimes P)\tilde{\xi} = 0$, which indicates that $\lim_{t\to\infty} \tilde{\xi} = 0$. Therefore, the observer in (7.6) can estimate $(-\mathcal{L}_1^{-1}\mathcal{L}_2 \otimes I)x_{\mathfrak{L}}$ under the designed event-triggered mechanism in Eq. (7.8). Further, we can see that the adaptive weights c_i ($i \in I[O+1, N]$) converge to the finite values.

Subsequently, we show that the multiagent system does not exhibit the Zeno behavior under the proposed event-triggered mechanism in Eq. (7.8). Here the contradiction argument is employed, without loss of generality, suppose that follower i exhibits the Zeno behavior. That is, $\lim_{k\to\infty} t_k^i = t_\infty^i < \infty$, which indicates that $\lim_{k\to\infty} T_k^i = 0$ with $T_k^i = t_{k+1}^i - t_k^i$. According to the event-triggered mechanism in Eq. (7.8), at $t = t_{k+1}^i$, we can obtain that

$$d_i^2(t_{k+1}^i)\|K\tilde{v}_i(t_k^i)\|^2 \geq \mu_{h^1}(t_{k+1}^i) \quad (7.29a)$$

$$\bar{c}_i^2(t_{k+1}^i)\|Ke_{\tilde{v}_i}(t_{k+1}^i)\|^2 \geq \mu_{h^2}(t_{k+1}^i). \quad (7.29b)$$

According to Eq. (7.10), for $t \in [t_k^i, t_{k+1}^i)$, we have

$$\lim_{k\to\infty} d_i^2(t_{k+1}^i) = \lim_{k\to\infty} (b_i(t_{k+1}^i) - b_i(t_k^i))^2 = 0 \quad (7.30a)$$

$$\lim_{k\to\infty} \|Ke_{v_i}(t_{k+1}^i)\|^2 = \lim_{k\to\infty} \|K(\tilde{v}_i(t_k^i) - \tilde{v}_i(t_{k+1}^i))\|^2 = 0. \quad (7.30b)$$

It follows from Eqs. (7.29) and (7.30) that the following inequalities can be obtained:

$$0 \geq \lim_{k\to\infty} \mu_{h^1}(t_{k+1}^i) = \mu_{h^1}(t_\infty^i) > 0 \quad (7.31a)$$

$$0 \geq \lim_{k\to\infty} \mu_{h^2}(t_{k+1}^i) = \mu_{h^2}(t_\infty^i) > 0 \quad (7.31b)$$

which is obviously contradictory. Therefore, the assumption is wrong that Zeno behavior does not occur in the multiagent system under the proposed event-triggered mechanism. The proof is completed. □

Given the time-varying formation vector $h = \text{col}(h_1, \cdots, h_O)$, where h_i is piecewise bounded continuously differentiable and satisfies that

$$\dot{h}_i = A_h h_i, \; i \in I[1, N] \quad (7.32)$$

and $A_h \in R^{n\times n}$ satisfies that

$$\begin{aligned}\Pi_{hi}A_h &= A_i\Pi_{hi} + B_i\Gamma_{hi}\\ C_0 &= C_i\Pi_{hi},\ i\in I[1, N+M].\end{aligned}\tag{7.33}$$

Subsequently, design the control law for the leader i $(i \in I[1, N]))$ as follows:

$$\begin{aligned}u_i &= K_{1i}\hat{\upsilon}_i + K_{2i}\upsilon_i + K_{3i}h_i\\ \dot{\hat{\upsilon}}_i &= A_i\hat{\upsilon}_i + B_iu_i + S_i(y_i - C_i\hat{\upsilon}_i),\ i\in I[1, N]\end{aligned}\tag{7.34}$$

in which $\hat{\upsilon}_i$ is the observer state and is used to estimate x_i based on the output information for leader i $(i \in I[1, N])$.

The following theorem shows that the formation leaders can achieve the predesigned formation and track the virtual leader simultaneously.

Theorem 7.1 *Under the condition of Assumptions 7.1 and 7.2, the time-varying output formation tracking problem for all leaders can be achieved under the designed observers in Eqs.* (7.5) *and* (7.6) *and the control laws in Eq.* (7.34), *if the feedback gain matrices are selected by* $A_i - S_iC_i$, $A_i + B_iK_{1i}$ *being Hurwitz matrices,* $K_{2i} = \Gamma_i - K_{1i}\Pi_i$, $K_{3i} = \Gamma_{hi} - K_{1i}\Pi_{hi}$, *in which* (Π_i, Γ_i) *and* (Π_{hi}, Γ_{hi}) *satisfy Eqs.* (7.5) *and* (7.34), *respectively.*

Proof Define the observer error in Eq. (7.34) as $e_{\hat{\upsilon}_i} = \hat{\upsilon}_i - x_i,\ i \in I[1, N]$. Calculate the derivative of $e_{\hat{\upsilon}_i}$ as

$$\dot{e}_{\hat{\upsilon}_i} = (A_i - S_iC_i)e_{\hat{\upsilon}_i}.$$

It follows from above formula that we can conclude that $\lim_{t\to\infty} e_{\hat{\upsilon}_i} = 0,\ i \in I[1, N]$ since $A_i - S_iC_i$ is Hurwitz.

Define $A = \text{diag}\{A_1, \cdots, A_N\}$, $B = \text{diag}\{B_1, \cdots, B_N\}$, $C = \text{diag}\{C_1, \cdots, C_N\}$, $K_i = \text{diag}\{K_{i1} \cdots, K_{iN}\}$, $(i = 1, 2, 3)$, $\Pi_{\mathfrak{L}} = \text{diag}\{\Pi_1, \cdots, \Pi_N\}$, $\Pi_{\mathfrak{L}h} = \text{diag}\{\Pi_{h1}, \cdots, \Pi_{hN}\}$, $\Gamma_{\mathfrak{L}} = \text{diag}\{\Gamma_1, \cdots, \Gamma_N\}$, $\Gamma_{\mathfrak{L}h} = \text{diag}\{\Gamma_{h1}, \cdots, \Gamma_{hN}\}$, $h = \text{col}(h_1, \cdots, h_N)$, $e_{\hat{\upsilon}} = \text{col}(e_{\hat{\upsilon}_1}, \cdots, e_{\hat{\upsilon}_N})$, $\upsilon = \text{col}(\upsilon_1, \cdots, \upsilon_N)$, $x_{\mathfrak{L}} = \text{col}(x_1, \cdots, x_N)$, $x_{\mathfrak{F}} = \text{col}(x_{N+1}, \cdots, x_{N+M})$.

Set $\tilde{x}_{\mathfrak{L}} = x_{\mathfrak{L}} - \Pi_{\mathfrak{L}}(1 \otimes x_0) - \Pi_{\mathfrak{L}h}h$. Then, the derivative of $\tilde{x}_{\mathfrak{L}}$ can be calculated from Eqs. (7.1), (7.2), (7.5), (7.32)–(7.34),

$$\begin{aligned}\dot{\tilde{x}}_{\mathfrak{L}} &= (A + BK_1)(\tilde{x}_{\mathcal{L}} + \Pi_{\mathfrak{L}}(1\otimes x_0) + \Pi_{\mathfrak{L}h}h) + BK_1e_{\hat{\upsilon}} + BK_2\upsilon\\ &\quad + BK_3h - \Pi_{\mathfrak{L}}(1\otimes A_0x_0) - \Pi_{\mathfrak{L}h}(I\otimes A_h)h\\ &= (A + BK_1)\tilde{x}_{\mathfrak{L}} + BK_1\Pi_{\mathfrak{L}}(1\otimes x_0) + BK_1\Pi_{\mathfrak{L}h}h + BK_1e_{\hat{\upsilon}}\\ &\quad + BK_2\upsilon + BK_3h - B\Gamma_{\mathfrak{L}h}h - B\Gamma_{\mathfrak{L}}(1\otimes x_0).\end{aligned}\tag{7.35}$$

Substituting $K_{2i} = \Gamma_i - K_{1i}\Pi_i$, $K_{3i} = \Gamma_{hi} - K_{1i}\Pi_{hi}$ into Eq. (7.35) yields

$$\dot{\tilde{x}}_{\mathfrak{L}} = (A + BK_1)\tilde{x}_{\mathfrak{L}} + BK_1e_{\hat{\upsilon}} + BK_2(\upsilon - (1\otimes x_0)).\tag{7.36}$$

Based on the fact that $A_i + B_i K_{1i}$ is Hurwitz, $\lim_{t\to\infty} e_{\hat{v}} = 0$, and $\lim_{t\to\infty} v - (1 \otimes x_0) = 0$, inspired by Lemma 7.1, we obtain $\lim_{t\to\infty} \tilde{x}_{\mathfrak{L}} = 0$. Then, define the time-varying output formation tracking error for leader i as $e_{y_i} = y_i - y_0 - C_0 h_i,\ i \in I[1, N]$.

Based on Eqs. (7.4) and (7.33), we can obtain the following equation:

$$e_{y_i} = C_i x_i - C_i \Pi_i x_0 - C_i \Pi_{hi} h_i.$$

Define $e_{\mathfrak{L}y} = \mathrm{col}(e_{y_1}, \cdots, e_{y_N})$. Then, it is easy to verify that $e_{\mathfrak{L}y}$ satisfies the following equation:

$$\begin{aligned}
e_{\mathfrak{L}y} &= C x_{\mathfrak{L}} - C\Pi_{\mathfrak{L}}(1 \otimes x_0) - C\Pi_{\mathfrak{L}h} h \\
&= C(\tilde{x}_{\mathfrak{L}} + \Pi_{\mathfrak{L}}(1 \otimes x_0) + \Pi_{\mathfrak{L}h} h) - C\Pi_{\mathfrak{L}}(1 \otimes x_0) - C\Pi_{\mathfrak{L}h} h \\
&= C\tilde{x}_{\mathfrak{L}}
\end{aligned}$$

which indicates that $\lim_{t\to\infty} e_{\mathfrak{L}y} = 0$ since $\lim_{t\to\infty} \tilde{x}_{\mathfrak{L}} = 0$. In conclusion, the proposed observers in Eqs. (7.6) and (7.7) and controller in (7.35) solve the time-varying output formation tracking problem for leaders. □

Inspired by Jiang et al. (2019), we design the observers for each follower to estimate $(-\mathcal{L}_4^{-1}\mathcal{L}_3 \otimes I)(1 \otimes x_0)$ and $(-\mathcal{L}_4^{-1}\mathcal{L}_3 \otimes I)h$, respectively. Firstly, design the following fully distributed event-triggered observer for ith follower to estimate $(-\mathcal{L}_4^{-1}\mathcal{L}_3 \otimes I)(1 \otimes x_0)$:,

$$\begin{aligned}
\dot{w}_i &= A_0 w_i + c_i K \tilde{w}_i(t_k^i) \\
c_i &= \alpha a_i + \beta b_i(t_k^i),\ i \in I[N+1, N+M]
\end{aligned} \tag{7.37}$$

where w_i is the observer state, $\tilde{w}_i = \sum_{j\in I[1,N]} a_{ij}(w_i - v_j) + \sum_{j\in I[N+1,N+M]} a_{ij}(w_i - w_j)$, the feedback gain matrix K is given later. c_i is the adaptive coupling weight for follower i, $\alpha > 0,\ \beta > 0$ are given positive constants, $a_i \in R$ and $b_i \in R$ satisfy

$$\dot{a}_i = \gamma \|K\tilde{w}_i(t_k^i)\|^2,\ t \in [t_k^i, t_{k+1}^i) \tag{7.38a}$$

$$b_i = \tilde{w}_i^T Q \tilde{w}_i \tag{7.38b}$$

with $\gamma > 0\ (\in R)$ and $Q > 0$ being the positive definite matrix to be given.

Define the measurement error $e_{\tilde{w}_i} = \tilde{w}_i(t_k^i) - \tilde{w}_i,\ t \in [t_k^i, t_{k+1}^i)$. The triggering instants t_k^i depend on the event-triggered mechanism in Eq. (7.9), triggering functions are designed by

$$h_i^1 = d_i^2 \|K\tilde{w}_i(t_k^i)\|^2 - \theta \|K\tilde{w}_i\|^2 - \mu_{h^1} \tag{7.39a}$$

$$h_i^2 = c_i^2 \|K e_{\tilde{w}_i}\|^2 - \vartheta \|K\tilde{w}_i(t)\|^2 - \mu_{h^2} \tag{7.39b}$$

and $d_i, \bar{c}_i$ are given in Eq. (7.11).

Define the following variables, $w = \text{col}(w_{N+1}, \cdots, w_{N+M})$, $\tilde{w} = \text{col}(\tilde{w}_{N+1}, \cdots, \tilde{w}_{N+M})$, $e_{\tilde{w}} = \text{col}(e_{\tilde{w}_{N+1}}, \cdots, e_{\tilde{w}_{N+M}})$. It follows from (7.37) that

$$\dot{w} = (I \otimes A_0)w + (\mathcal{C} \otimes K)(e_{\tilde{w}} + \tilde{w}) \tag{7.40a}$$

$$\tilde{w} = (\mathcal{L}_3 \otimes I)v + (\mathcal{L}_4 \otimes I)w. \tag{7.40b}$$

Then, calculate the derivative of $\tilde{w}$ as

$$\begin{aligned}\dot{\tilde{w}} = &(I \otimes A_0)\tilde{w} + (\mathcal{L}_4\mathcal{C} \otimes K)(e_{\tilde{w}} + \tilde{w}) + (\Phi\mathcal{L}_3 \otimes I)\Lambda_1 e_V \\ &+ (\Phi\mathcal{L}_3 \otimes I)\Lambda_2\tilde{v}'\end{aligned} \tag{7.41}$$

in which $\Lambda_1 = \text{diag}\{I \otimes SC_0, \mathcal{C} \otimes K\}$, $\Lambda_2 = \text{diag}\{0, \mathcal{C} \otimes K\}$, $e_V = \text{col}(e_{V_2}, e_{V_1})$, $\tilde{v}' = \text{col}(0, \tilde{v})$ are given. Note that element 0 in Λ_2 is the matrix with compatible dimensions.

It follows from Lemma 7.1 that $\lim_{t\to\infty} v = 1_N \otimes x_0$. It follows from Lemma 1.3 that the matrix $\mathcal{L}_4$ is nonsingular. Therefore, we can obtain that $\lim_{t\to\infty} w = (-\mathcal{L}_4^{-1}\mathcal{L}_3 \otimes I)(1 \otimes x_0)$ if and only if $\lim_{t\to\infty} \tilde{w} = 0$.

Subsequently, we give a lemma illustrating the feasibility of the designed observer, namely $\lim_{t\to\infty} \tilde{w} = 0$.

Lemma 7.2 *Under the condition of Assumptions 7.1–7.3, the fully distributed event-triggered observer in Eq.* (7.37) *is feasibile, i.e.,* $\lim_{t\to\infty} w = (-\mathcal{L}_4^{-1}\mathcal{L}_3 \otimes I)(1 \otimes x_0)$, *and the adaptive weights* c_i $(i \in I[N+1, N+M])$ *converge to the finite values, if the feedback gain matrix* K *is selected by* $K = -Q$ *and* $Q > 0$ *satisfies Eq.* (7.14), *the initial weight* $a_i(0) > \sqrt{\vartheta}/\alpha$ *is selected. In addition, the multiagent system does not exhibit the Zeno bahavior under the event-triggered mechanism in Eq.* (7.10) *and triggering functions in Eq.* (7.39).

Proof Select the Lyapunov function as

$$V = \sum_{i\in II[N+1,N+M]} \tilde{v}_i(\alpha a_i b_i + \frac{1}{2}\beta b_i^2) + \sum_{i\in II[N+1,N+M]} \frac{\alpha^2}{2\beta}\tilde{v}_i(a_i - \bar{a}) \tag{7.42}$$

where $\tilde{v}_i$ are given in Lemma 1.4, and $\bar{a} \in R$ is the positive constant with $\frac{\alpha^2\gamma\bar{a}\varpi}{\beta(1+\varpi)} - \frac{2\gamma\vartheta}{\beta a_m} - \frac{\gamma\bar{a}\vartheta}{\beta a_m^2}$ being sufficiently large, and a_m, ϖ are the same as Lemma 7.1.

Then, calculate the derivative of V along Eq. (7.38)

$$\begin{aligned}\dot{V} = &\sum_{i\in II[N+1,N+M]} \tilde{v}_i\bar{c}_i\dot{b}_i + \frac{\alpha}{\beta}\sum_{i\in II[N+1,N+M]} \tilde{v}_i(\beta b_i + \alpha a_i - \alpha\bar{a})\dot{a}_i \\ = &\tilde{w}^T\left(\Xi\bar{\mathcal{C}} \otimes (A_0^T Q + QA_0)\right)\tilde{w} - 2\tilde{w}^T(\bar{\mathcal{C}}\Xi\mathcal{L}_4\mathcal{C} \otimes K^T K)(\tilde{w} + e_{\tilde{w}}) \\ &- 2\tilde{w}^T(\bar{\mathcal{C}}\Xi\mathcal{L}_3 \otimes K)\Lambda_1 e_V - 2w^T(\bar{\mathcal{C}}\Xi\mathcal{L}_3 \otimes K)\Lambda_2\tilde{v}'\end{aligned}$$

$$+\frac{\alpha\gamma}{\beta}\sum_{i\in I[N+1,N+M]}\tilde{v}_i(\bar{c}_i-\alpha\bar{a})\|K\tilde{w}_i(t_k^i)\|^2. \tag{7.43}$$

Similar to Eq. (7.27), we can obtain the following inequality:

$$\begin{aligned}\dot{V}\leq&\tilde{w}^T(\Xi\bar{C}\otimes(A_0^TQ+QA_0)-\frac{1}{2}\Xi\bar{C}\otimes K^TK)\tilde{w}+\frac{\beta^2M\mu_{h^1}}{\rho_2}\\&+(\frac{M}{\rho_1}+\rho_4)\mu_{h^2}+4e_V^T\Lambda_1^T(\mathcal{L}_3^T\Xi\bar{C}\mathcal{L}_3\otimes I)\Lambda_1e_V\\&+4\tilde{v}'^T\Lambda_2^T(\mathcal{L}_3^T\tilde{\Xi}\bar{C}\Phi\mathcal{L}_3\otimes I)\Lambda_2\tilde{v}_1'\\\leq&-\tilde{w}^T(\Xi\bar{C}\otimes P)\tilde{w}+\frac{\beta^2M\mu_{h^1}}{\rho_2}+(\frac{M}{\rho_1}+\rho_4)\mu_{h^2}\\&+4e_V^T\Lambda_1^T(\mathcal{L}_3^T\Xi\bar{C}\mathcal{L}_3\otimes I)\Lambda_1e_V\\&+4\tilde{v}'^T\Lambda_2^T(\mathcal{L}_3^T\tilde{\Xi}\bar{C}\mathcal{L}_3\otimes I)\Lambda_2\tilde{v}_1'\end{aligned}\tag{7.44}$$

where ρ_i $(i=1,2,4)$ has the same definition as in Lemma 7.1. Since $\lim_{t\to\infty}e_V=0$, $\lim_{t\to\infty}\tilde{v}_1'=0$ hold, the rest is similar to Lemma 7.1, and omit it here. The proof is completed. □

Another fully distributed event-triggered observer for each follower is designed to estimate $(-\mathcal{L}_4^{-1}\mathcal{L}_3\otimes I)h$,

$$\begin{aligned}\dot{\varrho}_i&=A_h\varrho_i+c_i'K\tilde{\varrho}_i(t_k^i)\\c_i'&=\alpha a_i+\beta b_i(t_k^i),\ i\in I[N+1,N+M]\end{aligned}\tag{7.45}$$

where $\tilde{\varrho}_i=\sum_{j\in I[1,N]}a_{ij}(\varrho_i-h_j)+\sum_{j\in I[N+1,N+M]}a_{ij}(\varrho_i-\varrho_j)$, the matrix K is to be given, c_i' is the adaptive weight, α, β are the positive constants, $a_i\in R$ and $b_i\in R$ satisfy

$$\dot{a}_i=\gamma\|K\tilde{\varrho}_i(t_k^i)\|^2,\ t\in[t_k^i,t_{k+1}^i) \tag{7.46a}$$

$$b_i=\tilde{\varrho}_i^TQ\tilde{\varrho}_i \tag{7.46b}$$

with $r>0$ $(\in R)$ and $Q>0$ being the positive definite matrix to be specified.

Define the measurement error $e_{\tilde{\varrho}_i}=\tilde{\varrho}_i(t_k^i)-\tilde{\varrho}_i,\ t\in[t_k^i,t_{k+1}^i)$. The triggering instants t_k^i depend on the triggering mechanism in Eq. (7.10), where the triggering functions are given by

$$\begin{aligned}h_i^1&=d_i^2\|K\tilde{\varrho}_i(t_k^i)\|^2-\theta\|K\tilde{\varrho}_i\|^2-\mu_{h^1},\\h_i^2&=\bar{c}_i^2\|Ke_{\tilde{\varrho}_i}\|^2-\vartheta\|K\tilde{\varrho}_i(t)\|^2-\mu_{h^2}\end{aligned}\tag{7.47}$$

with the same d_i, $\bar{c}_i$ being given in Eq. (7.12).

Define the following variable, $\varrho=\mathrm{col}(\varrho_{N+1},\cdots,\varrho_{N+M})$, $\tilde{\varrho}=\mathrm{col}(\tilde{\varrho}_{N+1},\cdots,\tilde{\varrho}_{N+M})$, $e_{\tilde{\varrho}}=\mathrm{col}(e_{\tilde{\varrho}_{N+1}},\cdots,e_{\tilde{\varrho}_{N+M}})$. $h=\mathrm{col}(h_1,\cdots,h_N)$. It follows from Eq. (7.45) that

$$\dot{\varrho} = (I \otimes A_h)\varrho + (C' \otimes K)(e_{\tilde{\varrho}} + \tilde{\varrho}) \tag{7.48a}$$

$$\tilde{\varrho} = (\mathcal{L}_3 \otimes I)h + (\mathcal{L}_4 \otimes I)\varrho \tag{7.48b}$$

where $C' = \text{diag}\{c'_{N+1} \cdots, c'_{N+M}\}$. Calculate the derivative of $\tilde{\varrho}$ as

$$\dot{\tilde{\varrho}} = (I \otimes A_h)\tilde{\varrho} + (\mathcal{L}_4 C' \otimes K)(e_{\tilde{\varrho}} + \tilde{\varrho}). \tag{7.49}$$

Based on *Lemma 1.3*, one can obtain that the matrix $\mathcal{L}_4$ is nonsingular. Therefore, $\lim_{t\to\infty} \varrho = -(\mathcal{L}_4^{-1}\mathcal{L}_3 \otimes I)h$ can be obtained if and only if $\lim_{t\to\infty} \tilde{\varrho} = 0$.

Subsequently, we give a lemma that illustrates the feasibility of the designed observer in Eq. (7.45), i.e., $\lim_{t\to\infty} \tilde{\varrho} = 0$.

Lemma 7.3 *Under the condition of Assumptions 7.1–7.3, the fully distributed event-triggered observer in Eq.* (7.45) *satisfies that* $\lim_{t\to\infty} \varrho = (-\mathcal{L}_4^{-1}\mathcal{L}_3 \otimes I)h$ *and the adaptive weights* c'_i $(i \in I[N+1, N+M])$ *converge to the finite values, if the feedback gain matrix* $K = -Q$ *is selected, the positive definite matrix* $Q > 0$ *satisfies the following equation:*

$$QA_h + A_h^T Q - Q^2 + P = 0$$

with $P > 0$ *being positive definite matrix, and the initial weight* $a_i(0) > \frac{\sqrt{\vartheta}}{\alpha}$ *is chosen. Moreover, the multiagent system does not exhibit the Zeno behavior under the proposed event-triggered mechanism in Eq.* (7.11) *and triggering functions in Eq.* (7.47).

Proof Select the Lyapunov function as

$$V = \sum_{i\in I[N+1,N+M]} \tilde{\nu}_i(\alpha a_i b_i + \frac{1}{2}\beta b_i^2) + \sum_{i\in I[N+1,N+M]} \frac{\alpha^2}{2\beta}\tilde{\nu}_i(a_i - \bar{a}) \tag{7.50}$$

where $\tilde{\nu}_i$ are given in Lemma 1.4, and $\bar{a} \in R$ is a positive constant to be given.

Calculate the derivative of V as

$$\begin{aligned}
\dot{V} =& \sum_{i\in I[N+1,N+M]} \tilde{\nu}_i \bar{c}_i \dot{b}_i + \frac{\alpha}{\beta} \sum_{i\in I[N+1,N+M]} \tilde{\nu}_i(\beta b_i + \alpha a_i - \alpha\bar{a})\dot{a}_i \\
=& \tilde{\varrho}^T(\tilde{\Xi}\bar{C} \otimes (QA_h + A_h^T Q))\tilde{\varrho} - 2\tilde{\varrho}^T(\bar{C}\tilde{\Xi}\mathcal{L}_1 C' \otimes K^T K)(\tilde{\varrho} + e_{\tilde{\varrho}}) \\
&+ \frac{\alpha\gamma}{\beta} \sum_{i\in I[N+1,N+M]} \tilde{\nu}_i(\bar{c}_i - \alpha\bar{a})\|K\tilde{\varrho}_i(t_k^i)\|^2.
\end{aligned} \tag{7.51}$$

The rest is similar to the proof of Lemma 7.1 and is omitted here due to space limitations. The proof is completed. □

Remark 7.2 The designed observers in Eqs. (7.37) and (7.45) are inspired by Jiang et al. (2019). Note that the approach in Jiang et al. (2019) required continuous communication between neighboring agents. In fact, the actual system is equipped with limited resources. In order to minimize the resource damage, the observers make further improvements. We propose the fully distributed event-triggered observers.

7.3.2 *Formation Containment Control Law Design for the Followers*

Similar to the design of the formation control law, design the following fully distributed formation containment control law for the ith follower as follows:

$$
\begin{aligned}
u_i &= K_{1i}\hat{w}_i + K_{2i}w_i + K_{3i}\varrho_i \\
\dot{\hat{w}}_i &= A_i\hat{w}_i + B_iu_i + S_i(y_i - C_i\hat{w}_i),\ i \in I[N+1, N+M]
\end{aligned}
\tag{7.52}
$$

where $\hat{w}_i$ is the observer state to estimate x_i used the output information of follower i $(i \in I[N+1, N+M])$.

The following theorem ensures that the agents can achieve the formation containment problem.

Theorem 7.2 *Under the condition of Assumptions 7.1–7.3, the time-varying output formation containment problem of multiagent system composed of Eqs.* (7.1) *and* (7.2) *is solved by the proposed fully distributed controller in Eq.* (7.52), *if the selected feedback gain matrices satisfy that the matrices* $A_i - S_iC_i$, $A_i + B_iK_{1i}$ *are Hurwitz,* $K_{2i} = \Gamma_i - K_{1i}\Pi_i$, $K_{3i} = \Gamma_{hi} - K_{1i}\Pi_{hi}$, *where* (Π_i, Γ_i) *and* (Π_{hi}, Γ_{hi}) *satisfy Eqs.* (7.5) *and* (7.33), *respectively.*

Proof Construct the observer error in Eq. (7.52) as $e_{\hat{w}_i} = \hat{w}_i - x_i,\ i \in I[N+1, N+M]$. Then, calculate the derivatives of $e_{\hat{w}_i}$ as

$$
\dot{e}_{\hat{w}_i} = (A_i - S_iC_i)e_{\hat{w}_i}.
$$

Since the matrix $A_i - S_iC_i$ is Hurwitz, we can obtain that $\lim_{t\to\infty} e_{\hat{w}_i} = 0$.

Define the following variables: $\bar{A} = \text{diag}\{A_{N+1}, \cdots, A_{N+M}\}$ and the same to $\bar{B}$, $\bar{C}$, $\bar{K}_1$, $\bar{K}_2$, $\bar{K}_3$, $\Pi_{\mathfrak{F}} = \text{diag}\{\Pi_{N+1}, \cdots, \Pi_{N+M}\}$, $\Pi_{\mathfrak{F}h} = \text{diag}\{\Pi_{h(N+1)}, \cdots, \Pi_{h(N+M)}\}$, and the same to $\Gamma_{\mathfrak{F}}$ and $\Gamma_{\mathfrak{F}h}$, $e_{\hat{w}} = \text{col}(e_{\hat{w}_{N+1}}, \cdots, e_{\hat{w}_{N+M}})$, $\hat{w} = \text{col}(\hat{w}_{N+1}, \cdots, \hat{w}_{N+M})$, $\tilde{x}_{\mathfrak{F}} = x_{\mathfrak{F}} - \Pi_{\mathfrak{F}}(-\mathcal{L}_4^{-1}\mathcal{L}_3 \otimes I)(1 \otimes x_0) - \Pi_{\mathfrak{F}h}(-\mathcal{L}_4^{-1}\mathcal{L}_3 \otimes I)h$.

It follows from system dynamics in Eq. (7.2) and formation containment control law in Eq. (7.52) that

$$
\dot{x}_{\mathfrak{F}} = (\bar{A} + \bar{B}\bar{K}_1)x_{\mathfrak{F}} + \bar{B}\bar{K}_1e_{\hat{w}} + \bar{B}\bar{K}_2w + \bar{B}\bar{K}_3\varrho. \tag{7.53}
$$

In light of Eqs. (7.4), (7.33), and (7.53), we can obtain that

$$\begin{aligned}\dot{\tilde{x}}_{\mathfrak{F}} &= (\bar{A} + \bar{B}\bar{K}_1)\tilde{x}_{\mathfrak{F}} + \bar{B}\bar{K}_1\Pi_{\mathfrak{F}}(-\mathcal{L}_4^{-1}\mathcal{L}_3 \otimes I)(1 \otimes x_0)\\ &+ \bar{B}\bar{K}_1\Pi_{\mathfrak{F}h}(-\mathcal{L}_4^{-1}\mathcal{L}_3 \otimes I)h + \bar{B}\bar{K}_1 e_{\hat{w}} + \bar{B}\bar{K}_2 w + \bar{B}\bar{K}_3\varrho\\ &- \bar{B}\Gamma_{\mathfrak{F}}(-\mathcal{L}_4^{-1}\mathcal{L}_3 \otimes I)(1 \otimes x_0)\\ &- \bar{B}\Gamma_{\mathfrak{F}h}(-\mathcal{L}_4^{-1}\mathcal{L}_3 \otimes I)h.\end{aligned} \tag{7.54}$$

Using the fact that $K_{2i} = \Gamma_i - K_{1i}\Pi_i,\ K_{3i} = \Gamma_{hi} - K_{1i}\Pi_{hi}$, we obtain the following equation:

$$\begin{aligned}\dot{\tilde{x}}_{\mathfrak{F}} &= (\bar{A} + \bar{B}\bar{K}_1)\tilde{x}_{\mathfrak{F}} + \bar{B}\bar{K}_1 e_{\hat{w}} + \bar{B}\bar{K}_2\big(w - (-\mathcal{L}_4^{-1}\mathcal{L}_3 \otimes I)(1 \otimes x_0)\big)\\ &+ \bar{B}\bar{K}_3\big(\varrho - (-\mathcal{L}_4^{-1}\mathcal{L}_3 \otimes I)h\big).\end{aligned} \tag{7.55}$$

Furthermore, based on Lemmas 7.1 and 7.2, one can obtain that $\lim_{t\to\infty} w = (-\mathcal{L}_4^{-1}\mathcal{L}_3 \otimes I)(1 \otimes x_0)$ and $\lim_{t\to\infty} \varrho = (-\mathcal{L}_4^{-1}\mathcal{L}_3 \otimes I)h$. Since the matrix $A_i + B_i K_{1i}$ is Hurwitz and $\lim_{t\to\infty} e_{\hat{w}} = 0$, based on Eq. (7.55), we obtain $\lim_{t\to\infty} \tilde{x}_{\mathfrak{F}} = 0$.

Define the containment error as $e_{y_i} = y_i - y,\ y \in \mathrm{Co}(y_{\mathfrak{L}}),\ i \in I[N+1, N+M]$. Let $e_{\mathfrak{F}y} = \mathrm{col}(e_{y_{N+1}}, \cdots, e_{y_{N+M}}),\ y_{\mathfrak{F}} = \mathrm{col}(y_{N+1}, \cdots, y_{N+M}),\ e_{\mathfrak{F}y} = y_{\mathfrak{F}} - (-\mathcal{L}_4^{-1}\mathcal{L}_3 \otimes I)y_{\mathfrak{L}}$. It follows from Theorem 7.1 that the time-varying output formation error $e_{\mathfrak{L}y}$ satisfies $e_{\mathfrak{L}y} = y_{\mathfrak{L}} - (I \otimes C_0)(1 \otimes x_0 + h)$ and $\lim_{t\to\infty} e_{\mathfrak{L}y} = 0$. Therefore, we can obtain the following equation, based on the fact that $C_i\Pi_i = C_0$ in Eq. (7.4) and $C_i\Pi_{hi} = C_0$ in Eq. (7.34),

$$\begin{aligned}e_{\mathfrak{F}y} &= y_{\mathfrak{F}} - (-\mathcal{L}_4^{-1}\mathcal{L}_3 \otimes I)\big(e_{\mathfrak{L}y} + (I \otimes C_0)(1 \otimes x_0 + h)\big)\\ &= \bar{C}\tilde{x}_{\mathfrak{F}} + \bar{C}\Pi_{\mathfrak{F}}(-\mathcal{L}_4^{-1}\mathcal{L}_3 \otimes I)(1 \otimes x_0) + \bar{C}\Pi_{\mathfrak{F}h}(-\mathcal{L}_4^{-1}\mathcal{L}_3 \otimes I)h\\ &- (-\mathcal{L}_4^{-1}\mathcal{L}_3 \otimes I)\big(e_{\mathfrak{L}y} + (I \otimes C_0)(1 \otimes x_0 + h)\big)\\ &= \bar{C}\tilde{x}_{\mathfrak{F}} - (-\mathcal{L}_4^{-1}\mathcal{L}_3 \otimes I)e_{\mathfrak{L}y}\end{aligned} \tag{7.56}$$

which implies that $\lim_{t\to\infty} e_{\mathfrak{F}y} = 0$. Therefore, the time-varying output formation control problem of the considered multiagent system is solved. The proof is completed. □

7.4 Simulation Example

Consider the heterogeneous linear multiagent system composed of Eqs. (7.1) and (7.2), the communication topology between them is shown in Fig. 7.1, where the node 0 denotes the virtual leader, the nodes $1-7$ denote the formation leader, and the nodes $8-10$ denote the followers. Clearly Assumptions 7.1–7.3 hold.

Fig. 7.1 The communication topology

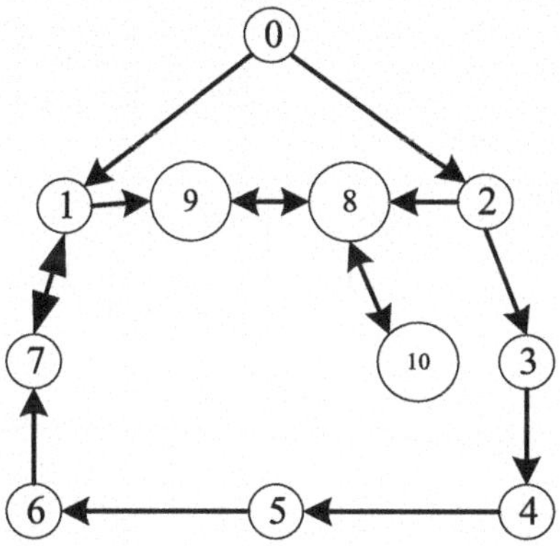

The system matrices in Eqs. (7.1) and (7.2) are given by $A_0 = \begin{bmatrix} 0 & 1 \\ -1 & 0 \end{bmatrix}$, $A_1 = A_2 = A_3 = A_8 = A_9 = \begin{bmatrix} 0 & 1 \\ 0 & -1 \end{bmatrix}$, $A_4 = A_5 = A_6 = A_7 = A_{10} = \begin{bmatrix} 0 & 1 & 0 \\ 0 & 1 & 1 \\ 1 & 0 & 1 \end{bmatrix}$, $B_1 = B_2 = B_3 = B_8 = B_9 = \begin{bmatrix} 1 & 1 \\ -2 & 2 \end{bmatrix}$, $B_4 = B_5 = B_6 = B_7 = B_{10} = \begin{bmatrix} 1 & 0 & 0 \\ 0 & 1 & 1 \\ 1 & 0 & 1 \end{bmatrix}$, $C_0 = \begin{bmatrix} 1 & 0 \\ 1 & 1 \end{bmatrix}$, $C_1 = C_2 = C_3 = C_8 = C_9 = \begin{bmatrix} 1 & 0 \\ 1 & -1 \end{bmatrix}$, $C_4 = C_5 = C_6 = C_7 = \begin{bmatrix} 1 & 0 & 0 \\ 0 & 1 & -1 \end{bmatrix}$.

The pole assignment method is utilized to select the eigenvalues of matrix $A_0 + SC_0$ as $-2,\ -5$, matrices $A_i - S_iC_i$ and $A_i + B_iK_{1i}$ $(i = 1, 2, 3, 8, 9)$ as $-2,\ -5$, and matrix $A_i - S_iC_i$ and $A_i + B_iK_{1i}$ $(i = 4, 5, 6, 7, 10)$ as $-2,\ -5,\ -8$.

According to the actual mission requirements, we let the formation leaders form the polygon formation, track on the virtual leader, and rotate. For this purpose, we design the time-varying formation vector h_i in Eq. (7.32) as

$$h_i = \begin{bmatrix} r\sin(\omega t + 2(i-1)\pi/7) + r\cos(\omega t + 2(i-1)\pi/7) \\ r\omega\sin(\omega t + 2(i-1)\pi/7) - r\omega\cos(\omega t + 2(i-1)\pi/7) \end{bmatrix}$$

in which $r = 1$, $\omega = 1$ are given. Furthermore, we can obtain that vector A_h satisfies $A_h = [0\ -w;\ w\ 0]$, and $\lim_{t\to\infty} \sum_{i\in I[1,N]} h_i = 0$.

Calculate the solution of regulator equations in Eqs. (7.4) and (7.33) as

$$\Pi_i = \begin{bmatrix} 1 & 0 \\ 0 & -1 \end{bmatrix}, \Gamma_i = \begin{bmatrix} -0.25 & 1.25 \\ 0.25 & 0.75 \end{bmatrix}, i = 1, 2, 3, 8, 9,$$

$$\Pi_{hi} = \begin{bmatrix} 1 & 0 \\ 0 & -1 \end{bmatrix}, \Gamma_{hi} = \begin{bmatrix} 0.25 & 0.25 \\ -0.25 & -0.25 \end{bmatrix},$$

and

$$\Pi_i = \begin{bmatrix} 1 & 0 \\ 0.4 & 1 \\ -0.6 & 0 \end{bmatrix}, \Gamma_i = \begin{bmatrix} -0.4 & 0 \\ 0 & 0 \\ -0.4 & 0.4 \end{bmatrix}, i = 4, 5, 6, 7, 10,$$

$$\Pi_{hi} = \begin{bmatrix} 1 & 0 \\ 0.8 & 0.2 \\ -0.2 & -0.8 \end{bmatrix}, \Gamma_{hi} = \begin{bmatrix} -0.8 & -1.2 \\ 0 & 0 \\ 0.4 & 0 \end{bmatrix}.$$

The initial state variables $x_i(0),\ i \in I[1, N+M]$, $v_i(0),\ i \in I[1, N]$, $\hat{v}_i(0),\ i \in I[N+1, N+M]$, $w_i(0),\ i \in I[1, N]$, $\hat{w}_i(0),\ i \in I[1, N]$, $\varrho_i(0),\ i \in I[1, N]$ are chosen randomly to be the interval (0, 1). Other relative parameters are chosen as $\alpha = 1,\ \beta = 10^{-4},\ \vartheta = 1,\ \gamma = 1,\ \theta = 1,\ \mu_{h^1} = \mu_{h^2} = 2e^{-0.5t}$. Then the simulation results are obtained as shown in Figs. 7.2, 7.3, 7.4, 7.5, 7.6, 7.7, and 7.8.

Figure 7.2 portrays the dynamic trajectory of the adaptive weights under Lemmas 7.1–7.3, from which it can be clearly observed that the values of the adaptive weights eventually stabilize and converge to finite values. Figure 7.3 depicts the event-triggered communication instants of individual agents, from which we can observe that multiagent system does not exhibit Zeno behavior under the proposed

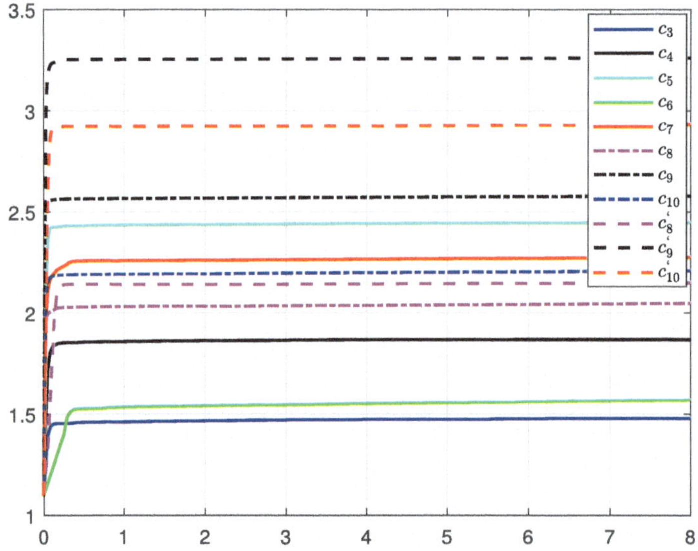

Fig. 7.2 The dynamic response of the adaptive weights $c_i(t)$, $c_i'(t)$ under Lemmas 7.1 and 7.3, reprinted from Zhang et al. (2023), copyright 2024, with permission from IEEE

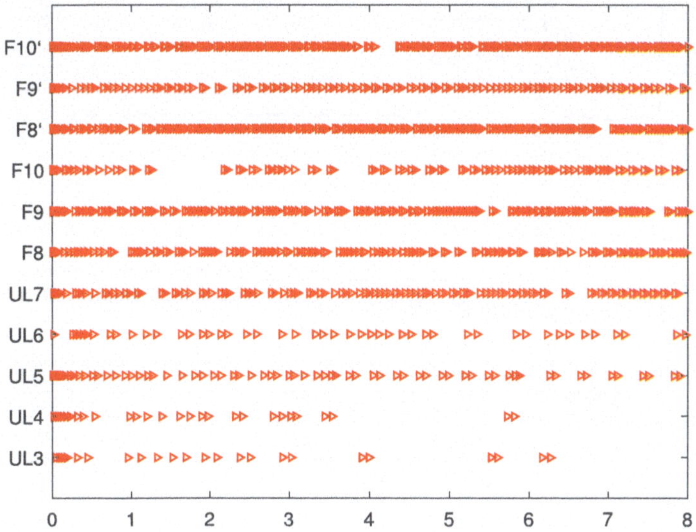

Fig. 7.3 The triggering instants under Lemmas 7.1 and 7.3, reprinted from Zhang et al. (2023), copyright 2024, with permission from IEEE

Fig. 7.4 The dynamic response of of $e_{v_i}(t)$ under Lemma 7.1, reprinted from Zhang et al. (2023), copyright 2024, with permission from IEEE

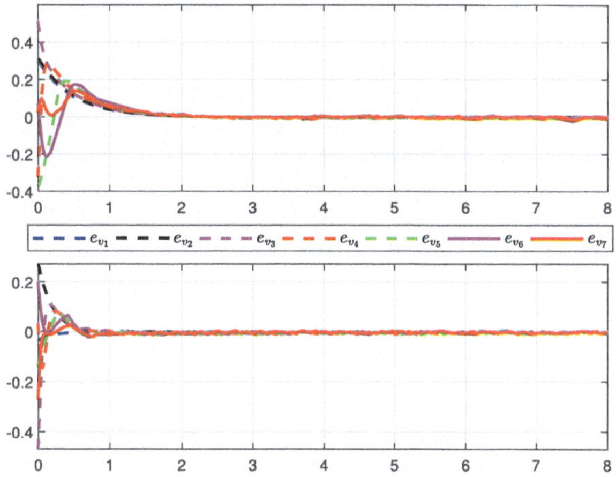

event-triggered mechanisms in this chapter. Figures 7.4, 7.5, 7.6, and 7.7 show the dynamic response of the observer errors e_{v_i}, $\tilde{w}_i$, $\tilde{\varrho}_i$ under Lemmas 7.1–7.3, from which we can see that the designed observers are feasible and the observer errors eventually converge to zero. Figure 7.9 shows the dynamic response of the formation error e_{γ_i} and Fig. 7.8 shows the dynamic response of the formation containment error e_{y_i}. Figures 7.8 and 7.9 indicate that the time-varying output formation containment control problem of heterogeneous linear multiagent system composed of Eqs. (7.1) and (7.2) is solved. Furthermore, from Fig. 7.9, one can observe that the predefined

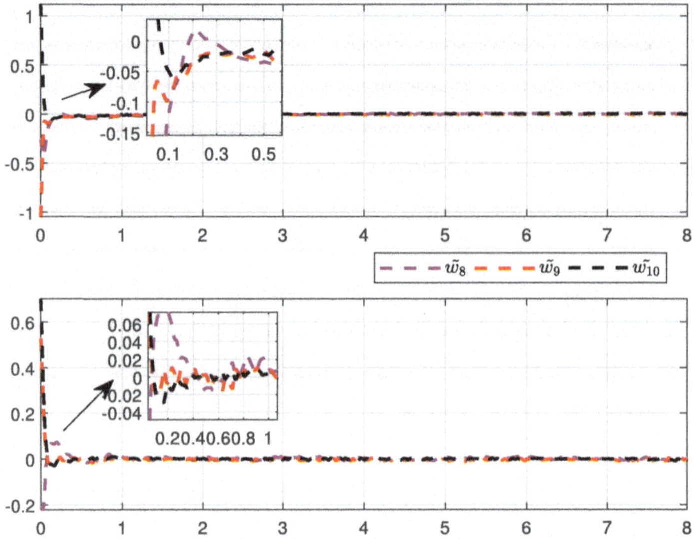

Fig. 7.5 The dynamic response of observer error $\tilde{w}_i(t)$ for each follower under Lemma 7.2, reprinted from Zhang et al. (2023), copyright 2024, with permission from IEEE

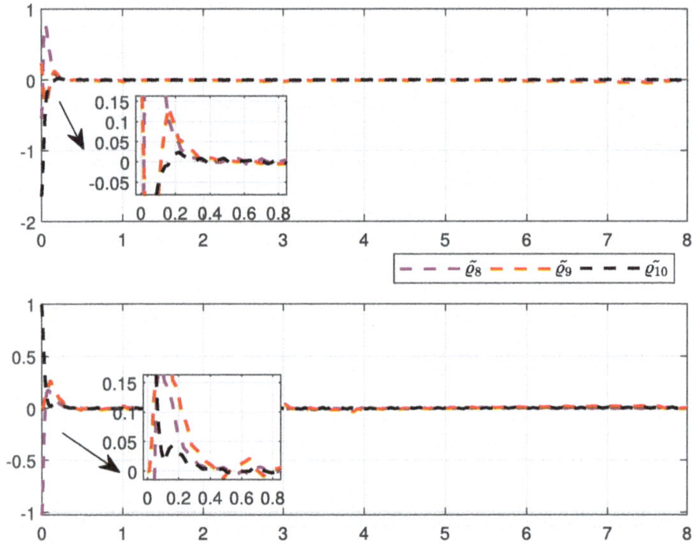

Fig. 7.6 The dynamic response of observer error $\tilde{\varrho}_i(t)$ for each follower under Lemma 7.3, reprinted from Zhang et al. (2023), copyright 2024, with permission from IEEE

formation is eventually realized, i.e., the formation leaders $1-7$ form a polygon and rotate around the virtual leader 0, represented by the red square, with the virtual leader located in the center of the formation, and the followers $8-10$ end up in the formation formed by the leader.

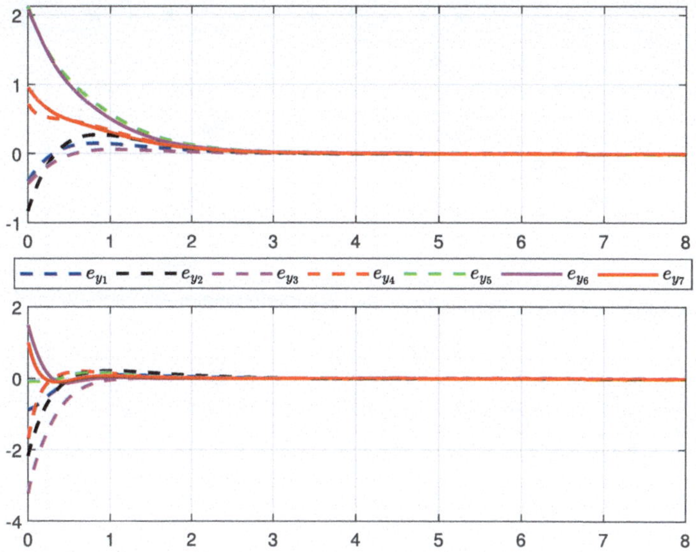

Fig. 7.7 The dynamic response of the formation error $e_{y_i} = y_i - y_0 - C_0 h_i,\ i \in I[1, N]$ under Theorem 7.1, reprinted from Zhang et al. (2023), copyright 2024, with permission from IEEE

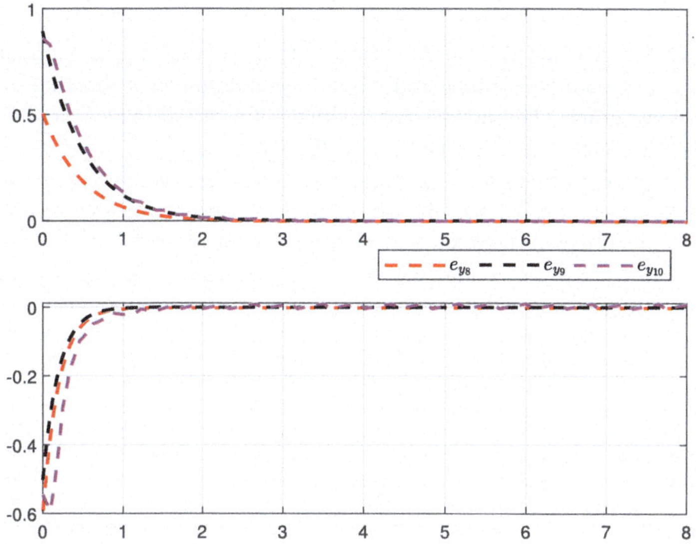

Fig. 7.8 The dynamic response of the formation containment error $e_{y_i} = y_i - y,\ y \in \mathrm{Co}(y_{\mathfrak{L}}),\ i \in I[N+1, N+M]$ under Theorem 7.2, reprinted from Zhang et al. (2023), copyright 2024, with permission from IEEE

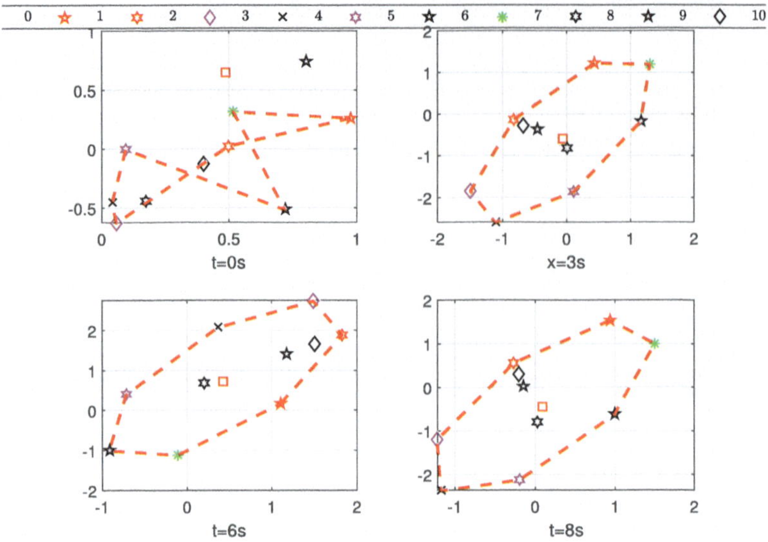

Fig. 7.9 Snapshots of all agents' outputs in different instants under Theorem 7.2

7.5 Summary

This chapter investigated the time-varying output formation containment control problem for heterogeneous linear multiagent systems under a directed communication topology graph. The agents are divided into virtual leaders and formation leaders, and followers based on tasks, and the dynamics between them is heterogeneous. Adaptive formation control protocol was first designed for the formation leader to enable it to realize the predefined formation and track on the virtual leader. Secondly, adaptive formation containment control protocol was designed to enable the followers to enter into the convex combinations formed by the leader. Finally, the effectiveness of the proposed method was verified by a numerical simulation.

References

Cheng B, Li Z (2019) Designing fully distributed adaptive event-triggered controllers for networked linear systems with matched uncertainties. IEEE Trans Neural Netw Learn Syst 30(12):3645–3655

Cheng B, Wu Z, Li Z (2021) Distributed edge-based event-triggered formation control. IEEE Trans Cybern 51(3):1241–1252

Dong X, Hu G (2017) Time-varying formation tracking for linear multiagent systems with multiple leaders. IEEE Trans Autom Control 3658–3664

Hu J, Bhowmick P, Lanzon A (2020) Distributed adaptive time-varying group formation tracking for multiagent systems with multiple leaders on directed graphs. IEEE Trans Control Netw Syst 7(1):140–150

Hua Y, Dong X, Wang J, Li Q, Zhang R (2018) Time-varying output formation tracking of heterogeneous linear multi-agent systems with multiple leaders and switching topologies. J Franklin Inst 356(1):539–560

Huang N, Sun Z, Anderson BD, Duan Z (2020) Distributed and adaptive triggering control for networked agents with linear dynamics. Inf Sci 517:297–314

Jiang W, Wen G, Peng Z, Huang T, Rahmani A (2019) Fully distributed formation-containment control of heterogeneous linear multiagent systems. IEEE Trans Autom Control 64(9):3889–3896

Jiang W, Rahmani A, Wen G (2021) Fully distributed time-varying formation-containment control for large-scale nonholonomic vehicles with an unknown real leader. Int J Control 94(4):1020–1032

Li Z, Wen G, Duan Z, Ren W (2015) Designing fully distributed consensus protocols for linear multi-agent systems with directed graphs. IEEE Trans Autom Control 60(4):1152–1157

Qian Y-Y, Liu L, Feng G (2021) Cooperative output regulation of linear multiagent systems: An event-triggered adaptive distributed observer approach. IEEE Trans Autom Control 66(2):833–840

Wang Y-W, Liu X-K, Xiao J-W, Lin X (2017) Output formation-containment of coupled heterogeneous linear systems under intermittent communication. J Franklin Inst 354(1):392–414

Zhang H, Cai Y, Wang Y, Su H (2020) Adaptive bipartite event-triggered output consensus of heterogeneous linear multiagent systems under fixed and switching topologies. IEEE Trans Neural Netw Learn Syst 31(11):4816–4830

Zhang J, Zhang H, Zhang K, Cai Y (2022) Observer-based output feedback event-triggered adaptive control for linear multiagent systems under switching topologies. IEEE Trans Neural Netw Learn Syst 33(12):7161–7171

Zhang J, Zhang H, Ming Z, Mu Y (2023) Adaptive event-triggered time-varying output bipartite formation containment of multiagent systems under directed graphs. IEEE Trans Neural Netw Learn Syst 34(11):8909–8922

Zuo S, Song Y, Lewis FL, Davoudi A (2019) Time-varying output formation containment of general linear homogeneous and heterogeneous multiagent systems. IEEE Trans Control Netw Syst 6(2):537–548

Chapter 8
Conclusion and Prospects

This book focuses on cooperative control problems of multiagent systems based on adaptive event-triggered methods. Cooperative control problems are categorized into consensus control, output regulate control, containment control, and formation control according to the given control objectives. When designing the control protocol, in order to avoid the dependence on the global information of communication topology, adaptive updating mechanism is designed to replace the fixed weights, and in order to reduce the communication frequency between intelligences, event-triggering mechanism is designed, and the design of controllers based on the adaptive mechanism and event-triggering mechanism is the problem of error closed-loop system. Through simulation experiments, the effectiveness, feasibility, and superiority of our method are verified. Specifically, the following content can be obtained:

1. For the undirected communication topology, the consensus problem of homogeneous linear multiagent systems is studied. Considering that it is difficult for each agent to obtain its own real state due to the complex communication environment, the model-based event-triggered model-based event-triggered observer is first used to estimate the state of each agent. Considering the limited communication resources of each agent and the large network of the multiagent system, a network of the multiagent system, a distributed adaptive event-triggered consensus protocol, and an adaptive protocol and an adaptive control law are designed by the state estimation, in which the dimension ofthe information exchanged between neighbors is the same as that of the system output, i.e., a low communication load is required. Therefore, the method in this chapter can effectively reduce the information transmission load of agents without increasing the network bandwidth.

2. For the undirected communication topology, the leader-follower consensus problem of a class of Lipschitz nonlinear multiagent systems is studied. First, the node-based and edge-based event-triggered mechanisms are designed, respectively, to the intermittent transmission of information from the leader to informed followers, and the intermittent asynchronous communication of agents. Second, static protocol

J. Zhang et al., *Adaptive Collaborative Control of Multiagent Systems by Event-Triggered Mechanisms*, Intelligent Control and Learning Systems 14,
https://doi.org/10.1007/978-981-97-7866-9_8

and an adaptive consensus protocol, where the design of the adaptive consensus of the adaptive consensus protocol is independent of the global information of the communication topology. The proposed method realizes intermittent asynchronous communication among agents, effectively avoids between agents, effectively avoids unnecessary information transmission, and thus saves communication resources.

3. For heterogeneous linear multiagent systems with DoS attacks and actuator faults, the output control problem is studied. The system matrix and the state of the leader can only be obtained by the subsystem directly connected to the leader. First, a distributed edge-based event-triggered observer is designed to estimate the system matrix and state of the leader. Using a lemma, the method of solving the regulator equation online under the condition that the system matrix of the leader is unknown is given. Secondly, under the influence of actuator faults, a distributed adaptive event-triggered fault-tolerant controller is proposed, which compensates for the influence of actuator faults on system stability and solves the output control problem. Edge-based event-driven detectors and event-triggered detectors are installed on the communication and controller sides, respectively, to ensure intermittent asynchronous communication and intermittent controller updates. The method in this chapter can not only ensure the safe operation of the network, but also recover the losses to a large extent when hacker attacks, and the system can still run stably when the actuators are faulty.

4. For the directed communication topology, the containment problem of linear multiagent systems is studied, where two cases of zero input leader and non-zero input leader are considered, respectively. Due to the asymmetry of the Laplace matrix, the method based on the adaptive mechanism to deal with the undirected communication topology graph is no longer feasible. For the zero input leader case, first, a distributed adaptive event-triggered control protocol and a composite event-triggered mechanism are constructed by using the relative information among agents. Second, a new computational method is proposed to ensure the measurement of relative information and intermittent communication among agents only at the triggering time. Finally, the proposed method is extended to the case of non-zero input leader. In addition, a comparative simulation is presented to show that the method proposed in this chapter has the advantages of fast convergence speed, low communication frequency of agents, and low update frequency of controllers.

5. For heterogeneous linear multiagent systems with disturbance, the time-varying output formation tracking problem is studied. Since not all followers can directly obtain the convex combination of the leader states, first, an adaptive distributed dynamic event-triggered compensator is designed for each follower to estimate the convex combination of the leader states, which achieves the purpose of saving communication resources. Second, this paper designs the time-varying output formation tracking controller based on the compensator and output information to realize the formation tracking problem of multiagent systems. In the simulation, comparative experiments are given to further illustrate the advantages of the methods in this chapter.

6. For the directed communication topology, the formation containment problems of heterogeneous linear multiagent systems are investigated. For formation containment problem, the dynamics of the leaders are heterogeneous and can be divided into three categories: virtual one, informed formation one, and uninformed formation one. First, two kinds of observers are respectively designed for informed and uninformed formation leaders to estimate the state of the virtual leader. Second, the formation control protocol is designed to make the formation leaders form an ideal formation. Finally, the containment control law is designed for followers to solve the formation containment problem. In this chapter, the dynamic formation function is introduced to avoid the collision between agents and obstacles, and the formation coverage area can be automatically adjusted according to the control requirements. Therefore, the methods in this chapter are more widely used.

This book investigates and explores the consensus problem, the output regulation problem, the containment problem, and the formation problem for multiagent systems. In summarizing the process of writing this book, the authors note that in future research work, further research may focus on the following areas.

The manufacturer's authorised representative in the EU is Springer Nature Customer Service Centre GmbH, Europaplatz 3, 69115 Heidelberg, Germany. If you have any concerns regarding our products, please contact ProductSafety@springernature.com

Printed and bound by CPI Group (UK) Ltd, Croydon, CR0 4YY

01/10/2026

02225966-0003